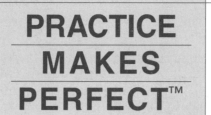

# PRACTICE
# MAKES
# PERFECT™

T0188531

# Biology

PRACTICE MAKES PERFECT™

# Biology

## THIRD EDITION

## Nichole Vivion

New York   Chicago   San Francisco   Athens   London   Madrid
Mexico City   Milan   New Delhi   Singapore   Sydney   Toronto

1 2 3 4 5 6 7 8 9   LON   27 26 25 24 23 22

ISBN        978-1-264-87494-1
MHID        1-264-87494-4

e-ISBN      978-1-264-87680-8
e-MHID      1-264-87680-7

Illustration credits can be found on pages 243–244, which is to be considered an extension of this copyright page.

McGraw Hill books are available at special quantity discounts to use as premiums and sales promotions or for use in corporate training programs. To contact a representative, please visit the Contact Us page at www.mhprofessional.com

McGraw Hill is committed to making our products accessible to all learners. To learn more about the available support and accommodations we offer, please contact us at accessibility@mheducation.com. We also participate in the Access Text Network (www.accesstext.org), and ATN members may submit requests through ATN.

# Contents

# Introduction

Biology, the study of all forms of life on Earth, is arguably the most integrative of the natural sciences. From the most minute perspective, biology is concerned with the inner workings of cells and the specific molecules they utilize for energy and nutrients. In this way, biology is directly dependent upon the principles of chemistry and ultimately, the principles of physics. Taking instead a large-scale approach, biologists seek to understand the interactions between organisms that exist in a shared habitat. This helps to uncover the multitude of interdependencies that have evolved between individual organisms in a population and between different species in a biological community. And as renowned biologist Theodosius Dobzhansky noted, "Nothing in biology makes sense except in light of evolution." It is the single most unifying principle in biology and must be the lens through which all of biology is examined.

*Practice Makes Perfect: Biology* will help you build the fundamental knowledge and skills necessary for developing a thorough understanding of the subject. This book is designed to be used by a novice to the field or someone who seeks a thorough review of biological concepts first encountered some time ago. It is organized topically from the most microscopic level of biological organization to the most macroscopic. Each chapter focuses on one major topic and presents thorough explanations and illustrative examples. Then you are encouraged to test your understanding and apply the knowledge you have gained by completing the exercises presented at the end of each chapter.

These exercises are both objective in nature, as in the Multiple Choice and Labeling/Interpreting Diagrams sections, and more subjective in nature, as in the Short Answer and Thinking Thematically sections. Each chapter includes Vocabulary Building and For Further Investigation, a section that provides ideas about how to apply the chapter's concepts to your everyday life. An answer key for all exercises is included at the back of the book for helpful feedback on your progress.

Biology as a discipline has undergone something of a revolution over the past few decades. The subfield of biotechnology has made numerous significant and exciting advances, some leading to controversial techniques like cloning and stem cell therapies. More general technological advances in microscopy and molecular laboratory techniques coupled with ongoing exploration of previously unknown regions of our planet continue to expand the breadth and depth of topics under the realm of biology. For that reason, essential to one's success in mastering biology is the ability to make connections between seemingly unrelated concepts, in much the same way that biologists seek connections to find the common thread to unify the vast biodiversity present on Earth.

To help in this effort, we take a two-pronged approach. First, this book utilizes the following six major themes relevant throughout the discipline of biology to relate topics from various chapters and to help manage the vast details involved:

- form facilitates function
- energy and organization
- continuity and change
- regulation and feedback
- natural interdependence
- science methodologies and applications to society

You will be asked at the end of each chapter to identify concepts from that chapter that act as relevant illustrations of three specified themes. This will enable you to make important connections between ideas and provide you with a meaningful context to make complex ideas easier to grasp.

Second, this book focuses on development of a new vocabulary through identification of a set of common Greek and Latin roots. (This way, new words can be understood through etymology, *etymo-* = "true meaning" and *-logy* = "study of," rather than by simply memorizing a given definition from a glossary.) As you begin reading, you will immediately notice many important vocabulary terms presented in boldface. Each time a new term is introduced, you should consult the Appendix to identify any Greek or Latin roots present in the term and to record new words in the blank column marked "More Examples." Note that this is a thorough but not exhaustive list of roots, so you are encouraged to consult other resources as well. You should then consider how those roots can be used to define the term. Often, you will be asked to do that as part of your vocabulary-building exercises at the end of each chapter. The more consistently you approach vocabulary building, the better equipped you will be to provide your own definitions to new terms based on your recognition of previously encountered roots.

One last thing to keep in mind when studying biology is that life is often unpredictable and exceptions to the rule usually exist. These exceptions are neither points of weakness in the field nor typically limits of our understanding, but rather are usually representative of the genetic diversity within organisms, their resourcefulness and ability to adapt, and the unique and sometimes strange evolutionary solutions arrived at by organisms of all sorts to remain successful on a dynamic Earth.

Armed with the strategies and information presented in this book, you will have a solid and detailed understanding of the fundamentals of the study of life. As animals ourselves, both intimately connected to yet often so distant from the complex ecosystem we inhabit, what could be more relevant, interesting, and empowering for us all?

# Biology

# FOUNDATIONS OF LIFE

# The Study of Living Things

**Biology** has been defined as the scientific study of life, but more specifically, it is the examination of the structure and function of all living things, or **organisms**. The breadth of its lens is quite remarkable, spanning from the tiniest bacterial cells to large and complex plant and animal life—and even beyond, to the interactions of organisms with each other and within the limitations presented by the physical environments they inhabit.

## The Characteristics of Life

The set of basic traits that all organisms share and that sustain every organism's life processes are collectively termed the *characteristics of life*. We can use these characteristics to defend why a simple bacterial cell is considered a living organism, while a virus is technically not (see Figure 1.1). First, a bacterium is **cellular** in composition; it is made up of a single cell that is surrounded by a protective external border and possesses internal structures that are essential for life. Within this single cell, all of the chemical reactions that are necessary to operate the bacterium's **metabolism** take place; it likely breaks down organic matter it has consumed as food to extract useful energy to keep life going.

This organism might seek out its food by detecting chemicals in its environment and then engulfing that substance. This exemplifies every organism's ability to respond to **stimuli**, just as we humans do with our senses of sight, smell, and touch. If the bacterial cell in question determines that something is out of balance internally, it can respond and attempt to reestablish the normal, beneficial state. This allows the organism to carry out another essential task—maintenance of **homeostasis**, a stable set of internal conditions despite likely changes in the external

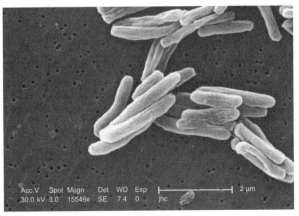

**Figure 1.1a** Bacterial Cells

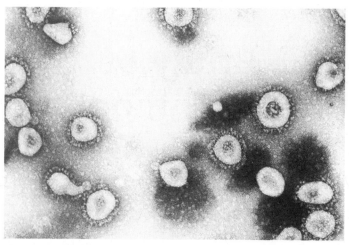

**Figure 1.1b** SARS-CoV-2, the virus that causes COVID-19.
https://en.wikipedia.org/wiki/Coronavirus#/media/File:TEM_of_coronavirus_OC43.jpg

environment. Failing to maintain homeostasis will cause the cell to use resources inefficiently, put it in a state of stress, and affect the quality of life.

As the bacterium continues to eat, it experiences **growth** by building up new structures and increasing the overall size of its cell. It eventually might undergo **reproduction** by copying its genetic instructions and splitting into two offspring cells. (Bacteria will be discussed in greater detail in Chapter 13.)

For comparison, a virus is far simpler in structure than even the least complex bacterium. A virus is composed of an assemblage of proteins and genetic instructions, so in that way, it is made up of some of the same structures as a typical cell. A key difference to note is that the proteins and genes in a virus are not arranged into a cellular form and actually appear more crystalline in nature. They are in fact so tiny as to only be seen with the most powerful electron microscopes that technology has to offer and are dwarfed in comparison to a typical bacterial cell. Additionally, viruses do not break down food for metabolic sustenance; they instead resort to the state of a true parasite and rely on the energy from the host cell to continue the infection. In fact, they rely on their host cell for energy, resources, and a place to replicate and make more viruses!

Although viruses certainly do mimic life in many ways, their inability to maintain any of these processes when outside of their host cell and their lack of any cellular structure place them in a realm somewhere between the living and the nonliving. (Viruses will be discussed in detail in Chapter 12.)

As you are certainly aware, life gets more complex than a typical bacterial cell. Most organisms that you are probably familiar with are unlike bacteria in that they are **multicellular**, composed of many cells (see Figure 1.2). This means that not only does each cell have to keep working appropriately, but that the cells must communicate with one another in order to keep the entire organism alive and well. It also means that when one of the cells divides, the organism itself is not necessarily reproducing. It is likely increasing its size in a way that is impossible for a **unicellular** bacterium—by simply adding more cells to its body.

The most complex of organisms have very well-organized structures. Cells that are part of the same structure and are working together for a common purpose are referred to as a **tissue**. In turn, different tissues work together in a structure to comprise an **organ**, and organs work together to carry out a common life function, comprising an **organ system**. All of the organ systems function together within the living organism. When considering all of the coordination

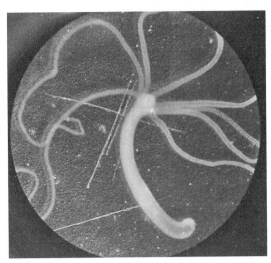

**Figure 1.2** Hydra, a Simple Multicellular Organism

necessary to maintain life, it becomes easy to see why organisms in nature are in a constant struggle for survival.

# The Biologist's Toolkit

Biologists, like all scientists, rely on the **scientific method** to guide their work. At its core, the scientific method is a set of steps that helps biologists organize their work and guide their scientific inquiry in a logical, meaningful way. One of its tenets is that any good scientific work should be replicable, able to be repeated by someone else if he or she follows the same steps using the same tools under the same set of conditions. In this way, the scientific method provides a way to validate one set of results and also to facilitate communication between scientists conducting research all over the globe.

The scientific method can take many specific forms depending on the type of research being conducted and the materials available for carrying out the work. In most cases, however, biological research should include a basic sequence of steps. First, careful **observations** are made about the living world, using the senses to describe natural phenomena. Questions are asked about what is seen, heard, and touched. For observations that can't be explained or aren't fully understood, a **hypothesis** is formulated. A hypothesis is a preliminary explanation for a question to be further explored and customarily includes a prediction, the presumed outcome assuming the hypothesis is eventually supported.

The hypothesis is then tested by designing and carrying out an **experiment**. In a **controlled experiment**, there are a minimum of two groups involved: the **experimental group** and the **control group**. The two groups are treated identically throughout the course of the experiment except for the variable being tested, which is only applied to the experimental group. This variable is called the **independent variable**. The control group then serves as a point of comparison; it demonstrates what would happen if the independent variable were not applied. If the test is being conducted in the field rather than the lab, a controlled experiment may be impossible or impractical. An **observational experiment** might instead be conducted to gather information about the subjects under natural conditions.

**Data** are collected and analyzed to generate the results of the experiment. These units of information can be **quantitative**, expressed in numbers, or **qualitative**, expressed in descriptive words. While many different types of data may be recorded, they will always include information on the **dependent variable**, the measure of the assumed effect of the independent variable. If there is any difference between the experimental and control groups, then it is assumed that the difference is due to the independent variable, the only known point of dissimilarity. When all of the evidence has been critically examined, **conclusions** are drawn, and the hypothesis is either supported or rejected. Note that a hypothesis is never really proven, leaving the scientific discipline open to new evidence, technologies, and ways of thought.

Sometimes **models** are made to visually or mathematically represent conclusions made in an experiment. Often results are published in scientific papers for research journals to communicate important findings worldwide, but only after a thorough and critical **peer review** process, which helps to identify any potential sources of bias or error in the design before a study is published.

Only very rarely are theories developed in biology. A **theory** is an overarching explanation for a major concept that is supported by vast quantities of data over time. In fact, the term *theory* is one of the most misunderstood in science. It is often misused by the general public, the media, and scientists themselves when the term *hypothesis* is actually intended. Recently, the misuse of the word *theory* has contributed to increased public confusion regarding the concept of evolution. The theory of evolution by means of natural selection is one of the most thoroughly supported ideas in all of science, documented with countless bits of evidence from all forms of life throughout its long history, but often is incorrectly dismissed as "just a theory." In biology, a theory holds great weight and should be considered as reliable as a scientific law.

One of the most revolutionary tools used by biologists to carry out important research is the **microscope**. First invented the late 1600s by Dutch spectacle makers, the **light microscope** uses light and one or more lenses to achieve **magnification**, an increase in the apparent size of an object. Adding other features to the microscope allowed for optimal **resolution**, or clarity of detail, to be obtained at various magnification levels. Microscopes are useful in viewing the tiniest life too small to be seen with the naked eye. They also help us examine the cellular structure of much larger organisms. Long after light microscopes had become a critical tool for biologists, **electron microscopes** were invented. The use of electrons much smaller than light photons for production of an image allowed for the first time for unicellular organisms to be observed in great detail and for viruses to be observed at all. **Scanning electron microscopes** produce very detailed surface images, while **transmission electron microscopes** allow for well-resolved cross sections of three-dimensional structures.

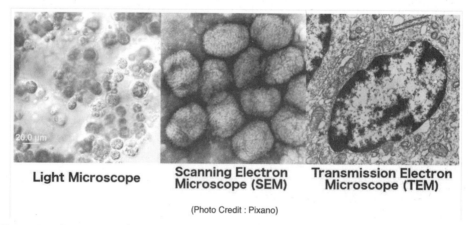

**Light Microscope**     **Scanning Electron Microscope (SEM)**     **Transmission Electron Microscope (TEM)**

(Photo Credit : Pixano)

**Figure 1.1c** Comparison of specimen under light and electron microscopes

Create a caption for Fig 1.1c by filling in the blanks: In the same relative field of view, light microscopes allow the viewer to see ...... while electron microscopes allow the viewer to see .....

**Vocabulary Building.** *Provide a definition for each of the following vocabulary terms. When possible, identify any roots in the term and use them to help create the definition.*

1. organism

2. homeostasis

3. metabolism

4. observation

5. controlled experiment

**Multiple Choice.** *Select the best response from the options provided to answer each question or to complete each statement.*

1. An organism achieves growth by
   a. breaking down food
   b. making new cells
   c. enlarging an existing cell
   d. both b and c

2. In a controlled experiment, researchers are trying to determine the effect of a new pharmaceutical drug on the body's ability to lose excess weight. Research subjects that do not receive the drug are
   a. part of the experimental group
   b. part of the control group
   c. part of both study groups
   d. not included in the study

3. In the same experiment described in #2, the weight of the subjects in both the control and experimental groups is the
   a. dependent variable
   b. hypothesis
   c. independent variable
   d. prediction

4. The scientific term that is most closely equated to "a law of nature" is
   a. model
   b. hypothesis
   c. theory
   d. prediction

5. If a microbiologist were examining the structure of the receptors on the outer surface of a bacterial cell, she should use a
   a. compound light microscope
   b. scanning electron microscope
   c. transmission electron microscope
   d. all of the above

**Short Answer.** *Write brief responses to the following questions.*

1. Use some of the characteristics of life to explain why mold growing on a damp wall is a living organism while frost accumulating in a freezer is not.

_____

_____

_____

_____

2. If a novice biologist forgot to include a control group in his experiment, how does this affect the conclusions he draws?

_____

_____

_____

_____

3. How are the concepts of magnification and resolution significant in using a microscope?

_____

_____

_____

_____

_____

**Interpreting Diagrams.** *Examine the following sketches produced by Robert Hooke, one of the first great microscopists. Look specifically for detail that reveals information about the levels of organization in biology. Answer the questions that follow.*

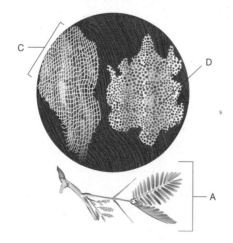

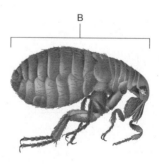

1. Which letter identifies a cell? _____

2. Which letter identifies a tissue? _____

3. Which letter identifies an organ? _____

4. Which letter identifies an organism? _____

**Thinking Thematically.** *For each of the following themes of biology, choose a different concept from this chapter and explain how it provides a useful illustration of that theme.*

1. regulation and feedback

_____

_____

_____

_____

_____

_____

2. science methodologies and applications to society

_____

_____

_____

_____

_____

_____

3. natural interdependence

_____

_____

_____

_____

_____

_____

## For Further Investigation

Read a news article or listen to a news report that showcases some new and interesting scientific research. Try to identify the type of experiment that was conducted and the hypothesis that was involved. What were the results of the study? Did the reporter or anyone quoted in the report use scientific method terminology correctly?

# Fundamentals of Chemistry

Underlying all of biology is **chemistry**, the study of the composition, properties, and behaviors of natural substances. As we have already established, all living things are cellular in structure and rely upon those cells to carefully contain and coordinate metabolic reactions constantly taking place within their boundaries. If the cellular chemistry is off-balance for the only cell in an unicellular organism, or for a critical mass of cells in a multicellular organism, then the organism itself is likely to become dysfunctional and could even die as a result. Chemistry is thus crucial to life. Two fundamental points of focus are significant for biology: the chemistry of water (detailed in this chapter) and the chemistry of carbon (organic chemistry and biochemistry; discussed in Chapter 3).

## Basic Chemical Principles

Before we can uncover the significance of the water molecule and the carbon atom, we must review some basic chemistry. Chemistry is the scientific study of **matter**, all things that have a mass and take up space. **Mass** is defined as the quantity of matter an object possesses (similar to the concept of **weight**, except that mass does not include the force of gravity on the object's mass, thus it doesn't change with different gravitational pulls), while the notion of space refers to the **volume** that an object's matter takes up.

At the simplest functional level, all matter is composed of structures called **atoms** (see Figure 2.1). An atom can be thought of as a (roughly) spherical space with a very small, dense core called the **nucleus**. Within the nucleus, positively charged **protons** and electrically neutral **neutrons** are held together by a very strong nuclear force.

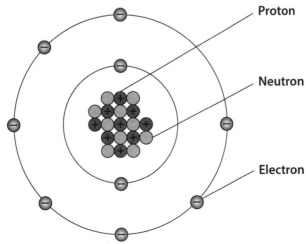

**Figure 2.1** Atomic Structure

Outside of the tiny atomic nucleus is a relatively huge space occupied by the third atomic particle, the **electron**. Electrons are negatively charged (–1) and move around within discrete regions known as **orbitals**. The electrons occupying the outermost orbitals of an atom are referred to as the **valence electrons**. An atom is considered neutral if the total number of protons and electrons are equal. If a neutral atom gains or loses electrons, then it becomes an **ion**, a charged particle. The loss of an electron creates a positive ion, called a **cation**, while the gain of an electron results in a negative ion, called an **anion**.

The number of protons present within an atom's nucleus uniquely identifies it as a specific element and is referred to as the element's **atomic number**. For example, the simplest element hydrogen (H) always has one proton in its nucleus and is thus given an atomic number of 1. Likewise, an examination of oxygen's nucleus will always reveal six protons, so it is assigned an atomic number of six. Although neutrons do not affect the charge of an atom as protons do, they do affect the mass of the atom. In fact, protons and neutrons are approximately equal in mass; the quantity of matter of a single particle of either is defined as one **atomic mass unit (amu)**. Knowing the atomic number of an atom and its experimentally determined **mass number**, the total number of protons and neutrons in an atom, one can deduce the number of neutrons that atom possesses.

Much of the information essential to understanding chemistry is summarized within a very useful tool called the **periodic table of elements**, commonly seen adorning the walls of any chemistry classroom. **Elements** are pure substances composed of only one type of atom. Every element known to science is included on the periodic table, along with its atomic number, mass number, and often other chemical properties. On the periodic table, the mass number is usually displayed with a decimal, seemingly indicating that there are fractions of particles present in an atom. What this actually indicates is that there are different **isotopes** that exist in nature, atoms of the same element but with differing numbers of neutrons. The mass number then is an average number based on the natural abundance of the different isotopes.

Most substances are not pure but are instead composed of more than one element and are considered compounds. A **compound** is created through the formation of a **chemical bond**, an energetic attraction between atoms that involves a pair of valence electrons. If the valence electrons from each atom involved in the bond are being shared, then the bond is referred to as covalent. **Covalent bonds** are strong and durable and create a new, higher level of structure called a **molecule**. Other types of bonds form because ions of opposite charges attract, creating instead an **ionic bond**. Ionic bonds are stronger than covalent ones but also can easily become disrupted in the presence of water or other similarly charged substances.

# Energy and Chemical Reactions

As already mentioned, energy is transferred when a chemical bond is created. Energy is defined as the ability to do work, and in this case, that involves the movement of electrons that contain mass. When a new substance is created due to the breaking or making of chemical bonds, a **chemical reaction** has occurred. In any chemical reaction, the starting substances are called **reactants**, and the end substances are called **products**. In order to generate products, the **activation energy** must be surmounted. To encourage reactions to occur without substantial amounts of energy invested, a **catalyst** might be added to speed up the reaction rate. Biological catalysts are called **enzymes**, the vast majority of which are protein in nature.

Regardless of whether enzymes are used and the reaction is catalyzed or is left alone uncatalyzed, the products of a chemical reaction may be at a higher or lower energy level than the reactants. If a reaction requires an energy investment to generate high-energy products, then the reaction is characterized as **endergonic**. When the opposite situation is true and energy is released during the reaction, it is considered **exergonic** (see Figure 2.2). Note that when heat **energy** is specifically involved, the terms **exothermic** and **endothermic** may be substituted.

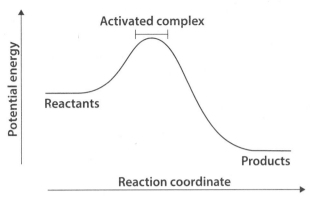

**Figure 2.2** An Exergonic Reaction

Create a caption for Figure 2.2 by using the terms *reactants*, *products*, and *activation energy* to explain how an exergonic reaction proceeds.

# Chemistry of Water

Understanding the chemistry of water is absolutely essential to understanding the structure and function of living things. Recall that nearly 75 percent of the Earth's surface is covered in water, so it contributes significantly to the basic makeup of the world's ecosystems, and approximately 70 percent of any given organism, such as a human, is composed of water. It is arguably the most significant inorganic molecule to all life, with oxygen likely coming in a close second (some organisms are obligate anaerobes, meaning that oxygen gas is toxic to them).

Water is an interesting molecule because of the unique interaction of its composite atoms. Water consists of a central oxygen atom covalently bonded to two hydrogen atoms. The electrons involved in the bond are not shared evenly between the oxygen and hydrogen atoms, however. Oxygen is known to be highly **electronegative**, having a particular attraction to electrons. Its pull is so strong that it usually pulls electrons away from the hydrogen atom, creating a **polar covalent** bond. Thus, the oxygen atom in any water molecule is described as being partially negative ($\delta^-$), and each hydrogen atom is described as being partially positive ($\delta^+$). The overall water molecule is characterized as **polar**, possessing internal areas of charge although the overall charge on the molecule is neutral.

When one polar water molecule interacts with another, the internal charges within one molecule attract to the oppositely charged regions of another water molecule. This creates a **hydrogen bond**, an intermolecular force between an electronegative atom and a positively charged hydrogen atom (see Figure 2.3). The hydrogen bonds that allow water to stick to itself

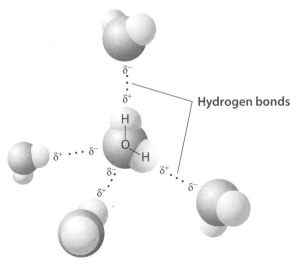

**Figure 2.3** Water and Hydrogen Bonding

contribute to the property of water known as **cohesion**, while the hydrogen bonds that form between water and another charged substance constitute its property of **adhesion**.

Other significant properties emerge for the water molecule, all based on its polarity and ability to hydrogen-bond. One such property is **capillarity**, the ability water has to travel up a narrow tube against gravity. Capillarity is really the collective force of cohesion and adhesion. Water sticks to itself to form a continuous water column (cohesion), and water sticks to the inside walls of the tube (adhesion), allowing it to "creep up" against the forces of gravity. A great example of capillarity is observed in plants as they passively move water up from their roots through the stems and toward the leaves in narrow tubes called *xylem*.

Water is also known for having high surface tension, allowing some insects to *walk on water*, and for atypically having a solid form (ice) that is less dense than the liquid form and floats on its surface. Relative to other substances, water has a **high specific heat**, meaning that it requires a good deal of energy to change its temperature. This again is due to hydrogen bonds that act as a thermal buffer for any body of water, preventing the water's temperature from fluctuating quickly and negatively affecting the organisms living within. When the ambient temperature begins to increase, the hydrogen bonds in the water must break before the water molecules can move about freely and actually contribute to an increase in temperature of the water. The reverse process occurs when cooling.

When two substances are mixed together evenly, a **solution** is created. A solution is always composed of a **solute** (often a solid) dissolved into a **solvent** (often a liquid). Most solutions relevant to biology, as you might imagine, are **aqueous solutions** in which the solvent is water. The quantity of solute dissolved into a given volume of solvent establishes the **concentration** of the solution.

In pure water, some of the molecules break apart into ions in a process called auto-ionization: one water molecule forms one hydroxide ion ($OH^-$) and one hydrogen cation ($H^+$) that then attaches to another water molecule, creating a hydronium ion ($H_3O^+$). Because the concentrations of both ions are equal in water, the substance is considered **neutral** on the **pH scale**. If hydronium ions are added to water, making their concentration higher than that of hydroxide, then the solution is considered **acidic**. When the hydroxide ions are instead higher in concentration than the hydronium ions, the solution is referred to as **basic** (or alkaline).

EXERCISE
**2·1**

**Vocabulary Building.** *Explain the relationship between the following pairs of vocabulary terms.*

1. atom, molecule

_____

_____

2. element, compound

_____

_____

3. polarity, hydrogen bond

_____

_____

EXERCISE
2·2

**Multiple Choice.** *Select the best response from the options provided to answer each question or to complete each statement.*

1. Protons are
   a. used to determine mass number
   b. positive in charge
   c. used to determine atomic number
   d. all of the above

2. If the number of neutrons changes within an atom, then what new form of that atom has been created?
   a. an ion
   b. an isotope
   c. an element
   d. a cation

3. Hydrogen gas ($H_2$) and oxygen gas ($O_2$) can be combined to make water ($H_2O$). In this situation, water is the
   a. product
   b. reactant
   c. enzyme
   d. substrate

4. The contents of the human stomach are very acidic, whereas the interior of the small intestine that receives the food from the stomach tends to be basic. Relative to that of the stomach, the pH of the small intestine should be
   a. higher
   b. lower
   c. equal
   d. neutral

5. Two pure substances are combined to make a third substance by sharing electrons evenly between them. The type of bond described is a
   a. hydrogen bond
   b. ionic bond
   c. polar covalent bond
   d. nonpolar covalent bond

EXERCISE
2·3

**Short Answer.** *Write brief responses to the following.*

1. The oxygen gas ($O_2$) that most organisms find necessary for life is made up of two identical oxygen atoms covalently bonded to one another. Does $O_2$ constitute a compound or a molecule? Explain.

_____

_____

_____
_____
_____
_____

2. A large drop of water forms from a leaky faucet before giving way to gravity and falling to the sink bottom. Explain the forces at work in forming and holding the drop.

_____
_____
_____
_____
_____
_____

3. Solid water is less dense than liquid water. Explain the ecological significance of such a property when considering a small pond freezing over for the winter.

_____
_____
_____
_____
_____
_____
_____

EXERCISE
2·4

**Interpreting Diagrams.** *Examine the following graph demonstrating the energy involved throughout the process of a reaction. Use the information in the graph to answer the questions that follow.*

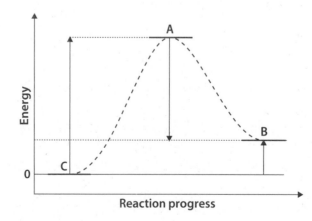

1. Which letter represents the reactants? _____

2. Which letter represents the products? _____

3. Which letter represents the activation energy? _____

4. Is the reaction endothermic or exothermic? Explain.

_____

**Thinking Thematically.** *For each of the themes of biology that follows, choose a different concept from this chapter and explain how it provides a useful illustration of that theme.*

1. form facilitates function

_____

_____

_____

_____

_____

_____

2. energy and organization

_____

_____

_____

_____

3. science methodologies and applications to society

_____

_____

_____

_____

_____

## For Further Investigation

For the next few days, pay special attention to the water around you. Why does ice float in your cold drink? Why does a cup of water form a convex dome on its surface before spilling over? Use a clear straw and glass half full with an aqueous solution of food coloring to demonstrate capillarity. Take a freshly cut white flower and place it in that same food color solution to change the color of the flower as the water moves up the stem and into the petals.

# Biochemistry

## Carbon and Organic Macromolecules

The foods you eat every day are most likely carbon-rich macromolecules in their various natural forms. **Proteins**, **carbohydrates**, and **lipids** (fats, oils, and waxes), constitute three of the four major macromolecule groups significant to living things. The fourth, **nucleic acids**, provides both the essential genetic instructions for life (**deoxyribonucleic acid**, or **DNA**) and the functional mechanisms for expressing genetic traits (**ribonucleic acid**, or **RNA**).

Although the exact structures vary markedly, all four types of macromolecules are **organic** in nature, composed of carbon atoms covalently bonded to other carbon atoms. If a molecule is a **hydrocarbon**, composed of only hydrogen atoms surrounding a carbon chain, then it is also considered **nonpolar**, having an even distribution of charge throughout, and **hydrophobic**, or water-fearing. Hydrophobic molecules like hydrocarbons repel water and other like substances. As small groups of atoms called **functional groups** are attached to basic hydrocarbons, these molecules take on different chemical characters and thus different functions. Functional groups are responsible for distinguishing proteins from carbohydrates at the fundamental level. Collectively, these macromolecules participate in the metabolism of a cell, which will be discussed in further detail in Chapter 5.

## Chemistry of Carbon

Even though water comprises most of the human body by mass, carbon is the fundamental building block of all of the body's cells and of the biological molecules that make up those cells. Carbon is so useful because it possesses four valence electrons, giving it the ability to form up to four bonds with other atoms. If carbon repetitively bonds to other carbon, long chains, branched chains, or even rings can be created (see Figure 3.1).

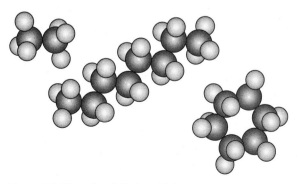

**Figure 3.1** Diversity of Carbon Molecules

Instead of forming four single bonds by sharing a single electron pair with each atom, a carbon atom might form a **double bond** by sharing two electron pairs with another atom. This leaves only two electrons available for other bonds, so either an additional double bond or two single bonds could form around the central carbon. Less frequently, three electron pairs can be shared between atoms to constitute a **triple bond**. That leaves the opportunity for only one single bond to form with the remaining electron. This extreme structural diversity of the carbon atom provides the foundation for the vast functional diversity necessary to support all life.

# Macromolecules and Metabolism

When animals like ourselves eat, we are most often consuming food on the macromolecule level; that is, we eat large, complex molecules referred to as **polymers** and made up of specific repetitive units. Through the digestive process, we are able to physically and chemically break down this polymer food into its simpler building blocks by breaking the chemical bonds that hold them together. Each building block, or **monomer**, is then used by our cells to carry out our metabolic needs and sustain life. Our bodies might use the monomers in different ways to build up a new polymer that they need, or they might use the monomer directly for cellular energy.

In order for a polymer to become digested into its monomer components, a **hydrolysis** reaction must take place (refer to Figure 3.2a). During this reaction, a water molecule is used to split two monomers apart. The $OH^-$ ion from $H_2O$ is added to one unit within the polymer, while the remaining $H^+$ ion from $H_2O$ is added to the adjacent unit. This helps to break the bond within the polymer and in turn releases one monomer.

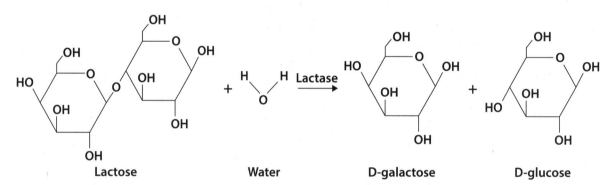

**Figure 3.2a** Hydrolysis of a Disaccharide

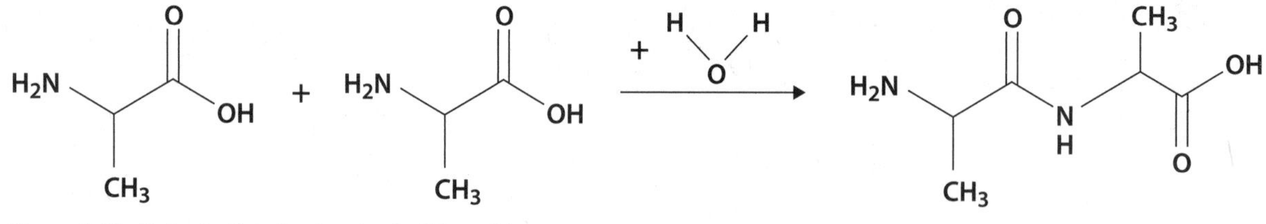

**Figure 3.2b** Dehydration Synthesis of a Dipeptide

Create a caption for Figure 3.2 to explain the difference between hydrolysis (a) and dehydration synthesis (b) reactions. Use the words *anabolic* and *catabolic* in your caption.

When, instead, monomers are being combined into a polymer, a reaction called a **dehydration synthesis** (or condensation reaction) takes place (see Figure 3.2b). Acting in the reverse mechanism as a hydrolysis reaction, a dehydration synthesis involves one monomer losing an $OH^-$ ion while another loses an $H^+$ ion. These ions are joined together to generate $H_2O$, and the two monomers then form a chemical bond, creating a **dimer**. If another dehydration synthesis occurs, then a third monomer would be added to the chain and the structure technically becomes a polymer.

# Structure and Function of the Macromolecules

When you consume proteins, they might be in the form of animal meat or legumes and soy. These polymer forms of proteins are called **polypeptides**. Polypeptides are composed of long chains of repeating monomer units named **amino acids** (see Figure 3.2) that are held together by special covalent attractions called **peptide bonds**.

Twenty different amino acids exist in nature; by combining them in different sequences into chains of different lengths, polypeptides serve many essential cellular roles. Proteins comprise the structural components of many animals' outer surfaces, like our own skin, hair, and nails. Protein fibers make up our muscles, too, and other cellular proteins act as channels, allowing cells to move molecules around. Many polypeptides serve as enzymes, biological catalysts that speed up reactions when present, but also allow for regulation of reactions by stopping a reaction when absent. When only two amino acids are bonded together into a structure called a **dipeptide**, they are often acting as small chemical messengers (hormones).

When you consume carbohydrates, you might be eating pasta, bread, or rice. These are complex carbohydrates, also called **polysaccharides**, because they are composed of many individual monomers, called **monosaccharides**, in branched form (see Figure 3.3). These types of carbohydrates are digested slowly and steadily and are thus a good source of intermediate-term energy for the body. That is why endurance athletes, like runners competing in a marathon, might "carbo-load" the night before a race by consuming complex carbohydrates.

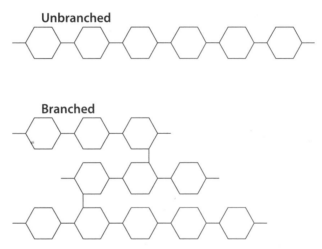

**Figure 3.3** Branched and Unbranched Polysaccharides

Often, we are instead consuming **simple sugars**, often in the **disaccharide** form (see Figure 3.2a). Table sugar, or sucrose, is one such simple sugar; all it takes is an enzyme to act on the sucrose molecule and break the **glycosidic linkage** that holds its two monomers together, and the individual monosaccharides are then free to be converted immediately into cellular energy (discussed in detail in Chapter 5). Consuming a lot of sucrose in one setting often leads to a burst of energy followed by a crash as the supply suddenly runs dry.

Lipids are slightly different from proteins and carbohydrates in that they are nonpolar. They don't have true monomers that can repetitively be bonded together in seemingly limitless ways and instead have very specific subtypes with characteristic structures. **Triglycerides**, the type of lipid humans typically consume if eating fat, are composed of three **fatty acid chains** connected to a single three-carbon molecule called **glycerol** (see Figure 3.4). Each fatty acid chain is long and hydrophobic, composed almost entirely of a chain of carbon atoms surrounded by hydrogen atoms. Through three dehydration syntheses, each fatty acid chain is linked up to a different carbon in the glycerol molecule, thereby forming a triglyceride.

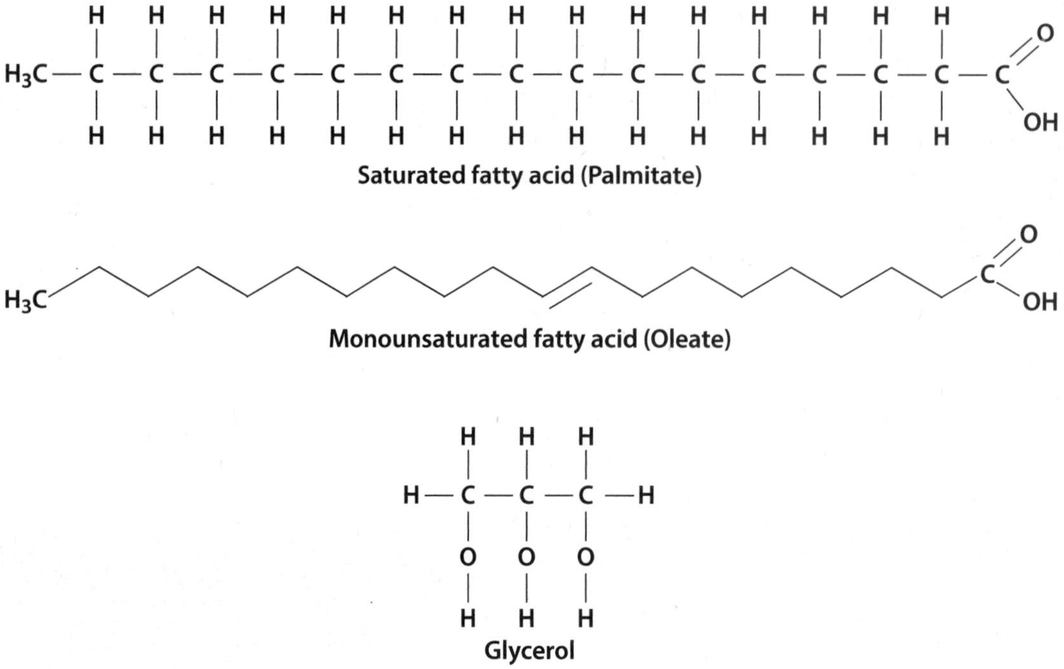

Saturated fatty acid (Palmitate)

Monounsaturated fatty acid (Oleate)

Glycerol

**Figure 3.4a** The building blocks of a triglyceride: glycerol and fatty acid chains.

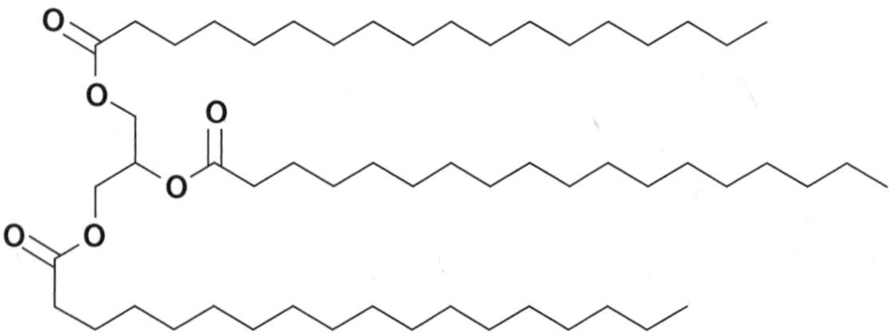

**Figure 3.4b** A Triglyceride

Some fatty acid chains are **saturated**, meaning that the long hydrocarbon chain is composed entirely of single bonds between carbons, and thus the carbons are saturated with hydrogen atoms surrounding them. Other fatty acids are instead **unsaturated**, containing at least one double bond between adjacent carbon atoms in the chain. The presence of a double bond reduces the opportunities for hydrogen to bond with carbon; therefore, the carbons are not saturated with the atom. From a health perspective, unsaturated fatty acids are preferable because they are easier for the body to process and are more likely to be broken down and used for energy. You may be familiar with the synthetic and somewhat controversial **trans fat**, a fat source created in a lab as the healthier unsaturated fat has its double bonds converted into single bonds, thereby becoming saturated.

Often, lipids are not consumed for energy like triglycerides but serve other functional roles for organisms. One ubiquitous lipid is the **phospholipid** that makes up the basic cell membrane structure of all organisms. A phospholipid is composed of a glycerol molecule attached to two fatty acid chains, but on the third carbon, a phosphate group is attached. This gives the phospholipid a "head" region that is polar and **hydrophilic** (water-loving) and a tail region that is highly **hydrophobic** (water-fearing). Having oppositional chemical character allows the phospholipid to act as a chemical screen for a cell (discussed further in Chapter 4).

## Deoxyribonucleic Acid (DNA)

| Sugar-phosphate backbone | Base pairs | Sugar-phosphate backbone |
|---|---|---|

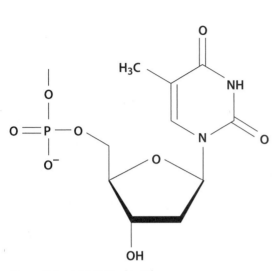

**Figure 3.5a** A DNA Nucleotide

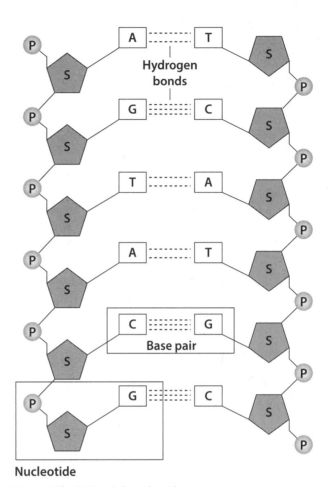

**Figure 3.5b** DNA, a Polynucleotide

Other lipids are either **waxes** that confer waterproofing to cells and surfaces or **steroids**, recognizable by their unique structure composed of four fused carbon rings. Many steroids are hormones for organisms, acting as chemical messengers between cells. Others are molecules like cholesterol—helpful when making up part of the typical cell membrane, but unhelpful when excess circulates in our bloodstream.

Nucleic acids, the final type of macromolecule, are long chains of repeating monomer units called **nucleotides**. Each nucleotide itself is actually a complex molecule, composed of a pentose sugar, a phosphate group, and a nitrogen-containing base (see Figure 3.5). All component parts are covalently bonded to each other, and nucleotides can be arranged into chains called **polynucleotides**.

Within a given nucleic acid like DNA or RNA, the sugar will always be the same in each nucleotide as will the phosphate group, but the particular nitrogenous base provides a point of distinction. There are four basic nitrogenous bases (referred to as A, T, C, and G) that together provide for the diversity required within the genetic code. Nucleic acids will be discussed in great detail in Chapter 7.

**Vocabulary Building.** *Explain the relationship between the following pairs of vocabulary terms.*

1. monomer, polymer

   _____

   _____

2. dehydration synthesis, hydrolysis

   _____

   _____

3. nucleic acid, nucleotide

   _____

   _____

**Multiple Choice.** *Select the best response from the options provided to answer each question or to complete each statement.*

1. Water is used as a reactant during which of the following types of reactions?
   a. dehydration synthesis
   b. hydrolysis
   c. condensation
   d. both a and c

2. Monosaccharides are to polysaccharides as amino acids are to
   a. nucleic acids
   b. dipeptides
   c. polynucleotides
   d. polypeptides

3. All of the following are considered lipids *except*
   a. waxes
   b. oils
   c. sugars
   d. fats

4. If a triathlete is preparing for a race and needs sustained energy, which of the following foods would be best to consume shortly before the competition?
   a. beef jerky for complete proteins
   b. fruit juice for simple sugars
   c. pasta for complex carbohydrates
   d. nuts for healthy fats

5. Which of the following macromolecules is responsible for storing the genetic code?
   a. DNA
   b. proteins
   c. RNA
   d. phospholipids

**Short Answer.** *Write brief responses to the following.*

1. How does the structure of the carbon atom influence the diversity of structures it can form? Explain.

   _____

   _____

   _____

   _____

   _____

2. Explain how using letters in the English alphabet to create different words is analogous to using monomers to build biological macromolecules.

   _____

   _____

   _____

   _____

   _____

3. Waxes are secreted on the outer surfaces of some organisms, like plants on their stems and leaves and mammals in their ears. What is the function of such a feature?

   _____

   _____

   _____

   _____

   _____

   _____

EXERCISE
3·4

**Labeling Diagrams.** *Fill in the blanks using the following terms to correctly label the diagram representing the parts of a nucleotide monomer.*

nitrogenous base

pentose sugar

phosphate group

1._____

2._____

3._____

**Thinking Thematically.** *For each of the following themes of biology, choose a concept from this chapter and explain how it provides a useful illustration of that theme.*

1. continuity and change

_____

_____

_____

_____

_____

_____

2. energy and organization

_____

_____

_____

_____

_____

_____

3. form facilitates function

_____

_____

_____

_____

_____

_____

_____

## For Further Investigation

Analyze some of the contents of your own kitchen for evidence of carbohydrates, proteins, and lipids. Scan the ingredients list on food labels for some now familiar-sounding names. Sugars often end in *-ose*, while proteins often have the root *pept-* in their names. Beware of fats listed as "partially hydrogenated oils," as those are the unhealthy trans fats!

# CELL BIOLOGY

# The Cell and Its Organelles

Recall that cells are the basic structural component of all living things, a fact easily observed when viewing plant or animal tissues under the microscope. Some organisms, like bacteria, are unicellular and thus must carry out all of their metabolic needs within the confines of a single cell. Other organisms are multicellular, having some cells that specialize in one metabolic process and other cells with different specialties, requiring coordination between cells.

The **cell theory**, an important theory in biology, describes the major roles that the cell plays in all organisms. First, the cell theory states that every organism is composed of at least one cell. All unicellular and multicellular organisms thus fit the bill. Second, cells are defined as the basic unit of structure and function in living things. If a cell is dysfunctional, then the organism may be (or will be, in the case of a unicellular organism). Finally, the cell theory states that all cells come from preexisting cells, a concept that will be discussed in detail in Chapter 6.

Viruses have already been described as existing somewhere between the living and nonliving. The cell theory makes this distinction fairly clear cut, however. Viruses are not cellular in structure and they cannot make other viruses without hijacking living cells. So, although they can wreak havoc on all types of living things from bacteria to plants and animals, viruses themselves are not considered true organisms.

## Basic Cell Features

Regardless of the type of cell under consideration, all cells have three basic parts: a cell membrane, cytoplasm, and DNA. The **cell membrane** (or **plasma membrane**) is a thin, flexible boundary that surrounds each individual cell. A main feature of the membrane is that it is **selectively permeable**, only allowing certain substances to cross and thus enter or leave the cell. In this way, the membrane helps maintain homeostasis for the cell. The structure and function of the membrane will be discussed further at the end of this chapter.

Just within the membrane lies the **cytoplasm**, an aqueous, jelly-like substance that suspends any internal structures and provides a solution within which some of the metabolic reactions of the cell take place. The aqueous portion of the cytoplasm is referred to as the **cytosol**. One structure found within the cytoplasm of every cell is the **ribosome**. These small, granular structures are the site of protein synthesis, so they contribute to the expression of the cell's genetic traits. Ribosomes are a complex structure composed of protein and nucleic acid molecules bound together. Finally, every cell also comes packaged with **DNA (deoxyribonucleic acid)**, the genetic instructions for the cell.

Structurally speaking, cells are classified as being one of two types: prokaryotic or eukaryotic. The simplest cells on Earth are called **prokaryotes** because their cell lacks a nucleus or any other internal compartments bound by their own membranes. More complex cells that possess internal, membrane-bound structures are instead known as **eukaryotes**. These internal structures, or **organelles**, provide the eukaryotic cell with internal areas of specialization. Consider the human body—each of us possesses organs that specialize in different tasks (e.g., the stomach specializes in digestion of proteins from food, while the kidneys help regulate water levels in the blood), but all must work together for the body to stay healthy. The cell is the same way. While it might possess various organelles each well suited for a given task, all of the organelles must be functioning properly for the cell to maintain homeostasis and, in doing so, a state of health for the cell.

## The Typical Animal Cell

In addition to the fundamental eukaryotic cell features already described, animal cells possess many other organelles that contribute to their overall functioning (see Figure 4.1). Animal cells typically have many **mitochondria**, organelles with double membranes that house important metabolic reactions and help generate cellular energy. Other structures called **lysosomes** contain **hydrolytic enzymes** within their membrane boundary. When lysosomes fuse with small containers of food that the cell has ingested, their enzymes help digest the food into monomers for eventual use as cellular energy.

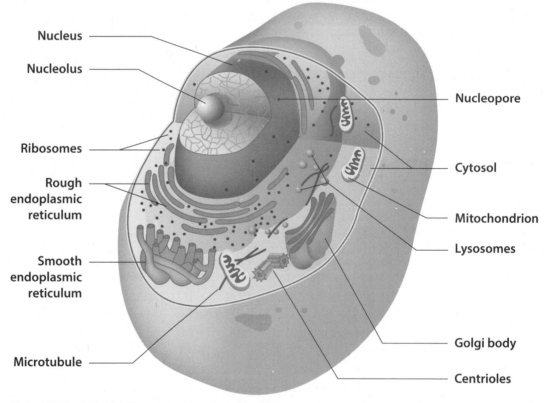

**Figure 4.1** An Animal Cell

Not only are those ubiquitous ribosomes scattered throughout the cytoplasm, many are also embedded within a membranous structure called the **endoplasmic reticulum (ER)** (see Figure 4.2). When ribosomes are found in the ER, it is then characterized as **rough ER** and is charged with packaging newly synthesized proteins into small, membrane-bound containers

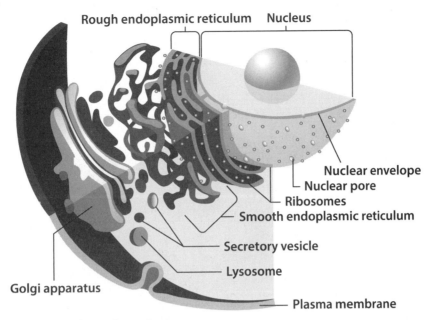

Rough endoplasmic reticulum    Nucleus

Nuclear envelope
Nuclear pore
Ribosomes
Smooth endoplasmic reticulum
Secretory vesicle
Lysosome
Golgi apparatus
Plasma membrane

**Figure 4.2** Endomembrane System

called **vesicles**. The membranes allow the vesicle to separate its contents from the cytosol and move it about or even out of the cell. These vesicles are temporary structures, meaning that they are created as membranes pinch off of existing organelles and are eventually engulfed by others as their membranes fuse and those contents move about and are chemically modified.

Adjacent regions to the rough ER but lacking ribosomes is the **smooth ER**. This portion synthesizes and packages lipids, and also assists in detoxification of the cytosol using specialized enzymes. Another organelle, the **Golgi body**, receives vesicles containing substances secreted from the ER and sorts, modifies, and packages them for export from the cell.

Animal cells are eukaryotic, so they possess a **nucleus** usually situated in the central region of the cell. The protective boundary of the nucleus is a double membrane called the **nuclear envelope**, but occasional **nuclear pores** provide for a means of exchanging substances between the nucleus and in the cell's cytoplasm. Usually, the DNA can be found within the nucleus in an uncondensed form called **chromatin**, giving the molecule a loose, threadlike appearance. When DNA is in chromatin form, another region within the nucleus called the **nucleolus** is also apparent as a dark, condensed core. The nucleolus is the site of the synthesis of **rRNA**, a structural component of ribosomes.

Just as many animals have a skeleton providing internal framework and support for the body, animal cells have a **cytoskeleton** composed of long, thin structures that come in two forms: microtubules and microfilaments. **Microtubules** are used throughout cells for many different purposes, as they make up **centrioles**, small, anchoring structures used in mitosis, and also external projections of the cell used for creating whiplike movements, called cilia and flagella. **Cilia** are shorter and more numerous, usually covering the entire outer surface of the cell. **Flagella** are longer than cilia and are positioned at one end of the cell. There may be a single flagellum present on a cell, or they might be grouped in pairs or triplets. **Microfilaments**—long, intertwined protein chains—help provide an internal framework for a cell and take part in cell movement and altering the shape of the membrane.

# The Typical Plant Cell

While plant cells and animal cells share most organelles in common, including the nucleus, there are important points of distinction. Most noticeably, plants have a thick, rigid **cell wall** exterior to the cell membrane (see Figure 4.3). The cell wall is composed of the fibrous polysaccharide

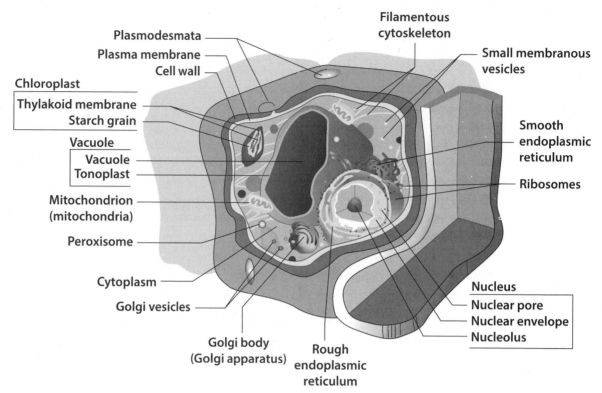

Plasmodesmata
Plasma membrane
Cell wall
Chloroplast
Thylakoid membrane
Starch grain
Vacuole
Vacuole
Tonoplast
Mitochondrion
(mitochondria)
Peroxisome
Cytoplasm
Golgi vesicles
Golgi body
(Golgi apparatus)
Rough
endoplasmic
reticulum
Filamentous
cytoskeleton
Small membranous
vesicles
Smooth
endoplasmic
reticulum
Ribosomes
Nucleus
Nuclear pore
Nuclear envelope
Nucleolus

**Figure 4.3** A Plant Cell

**cellulose**, known to humans in a dietary sense as fiber, which provides extra support to the cell and the collective plant body. In addition to most of the typical animal cell organelles, the plant cell also possesses **plastids** like the **chloroplast**. Plastids are storage compartments that usually house photosynthetic pigments for photosynthesis but can also store starch and other substances. Finally, plants also have a large central **vacuole** at their center, often displacing the nucleus off to one side. This central vacuole is essential in storing water and other nutrients for the cell, and in doing so, provides some aqueous "filling" to the cell.

It should be noted also that plants lack some structures typical to animals. Since plants make their own food, they have little need for the lysosomes that are so common and essential to animal cells. Additionally, they lack the centrioles that are used in animal cells in the nuclear division process called mitosis. There may be analogous structures in plants that simply have not yet been observed, or there may be other mechanisms at work that are not fully understood.

# Membrane Structure and Function

As has been established, all cells have an outer cell membrane. Due to its structure, the cell membrane is selectively permeable, allowing only certain substances to pass back and forth by creating a physical and chemical screen. Recall from Chapter 3 that the membrane is composed primarily of phospholipids. The head region of each phospholipid is **hydrophilic** (partially charged and thus capable of interacting with polar water molecules), while the tail region is nonpolar and instead **hydrophobic**. These individual macromolecules are arranged themselves into a **phospholipid bilayer** such that the heads are always facing out and the tails collectively face in and create the interior thickness of the membrane (see Figure 4.4).

Embedded within the phospholipid bilayer are different types of proteins and often other substances, like molecules of cholesterol that help stabilize the membrane and make it less permeable to water-soluble substances. Some proteins are **transmembrane**, spanning the entire

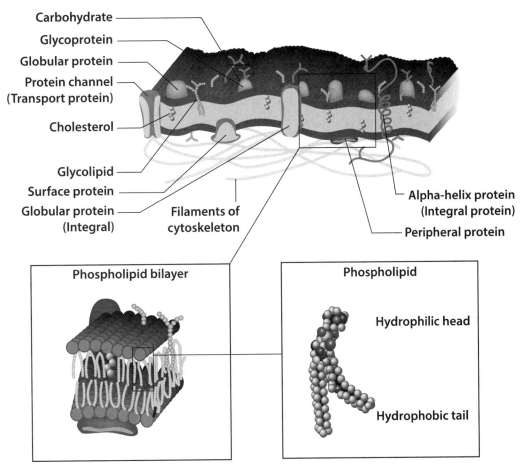

Cell membrane

Carbohydrate
Glycoprotein
Globular protein
Protein channel
(Transport protein)
Cholesterol
Glycolipid
Surface protein
Globular protein
(Integral)
Filaments of
cytoskeleton
Alpha-helix protein
(Integral protein)
Peripheral protein

Phospholipid bilayer

Phospholipid
Hydrophilic head
Hydrophobic tail

**Figure 4.4** Cell Membrane Structure

thickness of the membrane and having exposed regions on the exterior and interior surfaces of the membrane. Other proteins are **peripheral**, only studding one surface or the other and not spanning the thickness of the membrane.

The structure of the overall membrane helps the cell maintain homeostasis through its selective permeability. Most of the membrane is made up of a phospholipid bilayer that prevents the vast majority of substances from being transported across. The lipids themselves are closely packed, so only very small molecules, atoms, or ions could potentially pass. A substance must also meet certain chemical criteria, most importantly possessing very little or no charge that would be repelled by the thick hydrophobic interior. Thus, oxygen and carbon dioxide are the two main molecules present in and around most cells that can squeeze through the bilayer. Water, being polar, can only leak across very slowly through the bilayer directly, mainly due to its small size and high concentration. It also moves more freely through **aquaporins**, special pores in the cell membrane that permit rapid water movement.

Other substances that are too large and/or too charged to pass through the bilayer instead can rely on transmembrane channel proteins. All channel proteins are specialized to some degree to the substance(s) they transport, but some of these proteins are **gated channels** and more highly regulated, while others are always open.

Whether moving through the bilayer or utilizing a channel protein, a substance being transported across the membrane utilizes one of two main types of transport, passive or active.

# Passive Transport Mechanisms

In **passive transport**, no energy is invested by the cell because the movement is occurring down the **concentration gradient**. When molecules are in an area of high concentration within a larger space, they will bump into each other until they have spread out evenly, or reached **equilibrium**. The most basic type of passive transport is referred to as simple **diffusion**, as takes place when oxygen moves from the air sacs inside our lungs where the concentration is high into the oxygen-poor capillaries that surround them (see Figure 4.5).

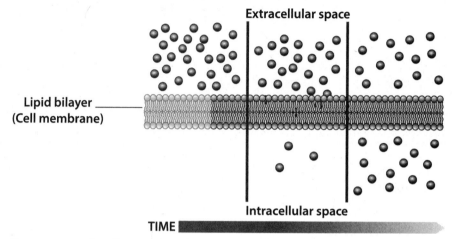

**Figure 4.5** Simple Diffusion Across a Phospholipid Bilayer

When water is the substance that is diffusing, the process is called **osmosis**. Osmosis occurs between two different solutions separated by a membrane when the solutions have different concentrations of solutes. A solution is described as **hypertonic** when it has a higher concentration than another solution. That other solution with a relatively low concentration would instead be described as **hypotonic**. Two solutions in equilibrium are characterized as **isotonic**, having equal concentrations.

It is important to note that, if solutes were permitted to move across the membrane, then they certainly would diffuse as well when a concentration gradient is present. But as we've already established, most substances can't diffuse across the bilayer. Most of the time, water is the substance that does most of the passive transport to achieve equilibrium. Water always moves passively down its gradient as well, but because water is the solvent and not the solute, water moves from a hypotonic solution to a hypertonic solution. Another way of thinking about it is that water naturally moves in a direction that results in dilution of the more concentrated side.

If water movement either into or out of an animal cell is unregulated or the cell is in an extreme environment, states of stress result for the cell. For example, if a cell is in a very salty environment, then the gradient dictates that water will move from the cytoplasm of the cell to the salt solution outside. Continued unchecked, this could result in extreme dehydration and wilting of the cell known as **plasmolysis**. In the reverse situation, cells flooded with distilled water find themselves in a hypotonic environment. Water floods the cytoplasm of the cell, diluting it and potentially giving rise to **cytolysis**, the bursting of the cell itself.

Some substances move passively into or out of cells but require the assistance of specific channel proteins to do so. This **facilitated diffusion** still moves the substance down its gradient from high to low concentration but would not take place were it not for the transport proteins that shield the moving substance from the bilayer.

# Active Transport Mechanisms

Many cells do not want to be in exact equilibrium with their surroundings and must then conduct active transport to achieve homeostasis. **Active transport** relies on the investment of energy by the cell because it moves substances against the concentration gradient. Energy is used to force a substance from an area of low concentration to an area of higher concentration. The form of energy that cells use to power such a process is typically in the form of **adenosine triphosphate**, or **ATP**. Discussed in detail in Chapter 5, ATP acts as energy when a phosphate group is broken off of the molecule and the bond energy is released.

One way that cells achieve such movement is through special transmembrane proteins called **cell membrane pumps** (see Figure 4.6). The classic example of such is the **sodium-potassium pump**, a protein found in high concentrations along the surfaces of nerve cells. Without the action of the sodium-potassium pump, which shuttles sodium ions ($Na^+$) out of the cell in exchange for potassium ions ($K^+$), nerves would not be able to transmit messages properly.

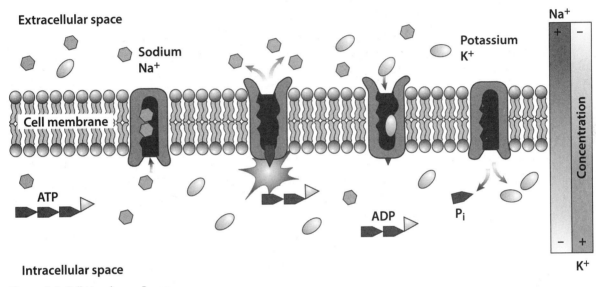

**Figure 4.6** Cell Membrane Pump

Create a caption for Figure 4.6 by first identifying as many structural components of the membrane as possible and then utilizing them to explain the mechanism of action in a cell membrane pump.

When the pump is open toward the inside of the cell, the shape of the protein has a high affinity for $Na^+$. Three sodium ions bind to the protein, but only when an ATP molecule hydrolyzes its terminal phosphate and attaches it to the protein does it have enough energy to change shape. This conformational change causes the protein to now be open to the outside of the cell and in a form with a low affinity for $Na^+$ but a high affinity for $K^+$. The three $Na^+$ ions thus break away from the protein and make way for two $K^+$ ions to bind on a different region. The phosphate group still bound to the protein (and previously donated from ATP) now is released from the pump, and this energy transfer returns the pump to its original shape. The protein now has little affinity for $K^+$, so those ions break free, and the cycle may begin anew.

A final type of active transport is called **bulk flow**, the movement of macromolecules or large volumes of a substance into or out of a cell. When the direction of movement proceeds from outside of the cell to the interior, the process is termed **endocytosis** (see Figure 4.7); when bulk flow is directed in the reverse direction, **exocytosis** instead takes place. If a cell moves fluids and dissolved solutes into the cell, it is conducting a specialized form of endocytosis called **pinocytosis**. Endocytosis of solid particles or of entire cells is referred to as **phagocytosis**.

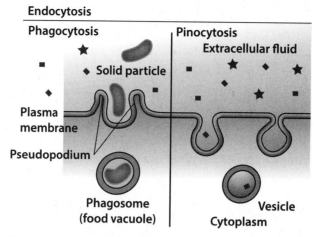

Endocytosis

Phagocytosis

Pinocytosis

Extracellular fluid

★ Solid particle

Plasma membrane

Pseudopodium

Phagosome (food vacuole)

Vesicle

Cytoplasm

**Figure 4.7** Types of Endocytosis

In any case of bulk flow, a vesicle is utilized to contain the substance in question while it is inside the cell. For example, when a cell engulfs organic matter for food, it does so by first manipulating its cell membrane so that it eventually surrounds the food. When the two sides of the membrane meet, they fuse and create the vesicle that now contains the food. This vesicle can fuse with a lysosome for further digestion if needed.

If a cell instead had wastes to get rid of, it would direct a vesicle of waste toward the cell membrane. When the membrane of the vesicle fuses with the membrane of the cell, the contents of the vesicle are released to the exterior of the cell. As you might imagine, orchestrating this much movement of the membrane requires ATP. The energy investment is worth it, though, if the cell is able to access nutrients, rid itself of wastes, and maintain overall homeostasis more effectively.

**EXERCISE**

**4·1**

**Vocabulary Building.** *Explain the relationship between the following pairs of vocabulary terms.*

1. cell membrane, selectively permeable

_____

_____

2. osmosis, concentration gradient

_____

_____

3. vesicle, phagocytosis

_____

_____

**Multiple Choice.** *Select the best response from the options provided to answer each question or to complete each statement.*

1. Any typical cell would be expected to have all of the following *except*
   a. ribosomes
   b. a cell membrane
   c. DNA in the nucleus
   d. cytoplasm

2. The primary function of the rough ER is to
   a. synthesize and package proteins
   b. produce cellular energy
   c. synthesize lipids
   d. copy the DNA

3. Which of the following would be expected in a typical animal cell but not in a plant cell?
   a. ribosomes
   b. mitochondria
   c. chloroplasts
   d. centrioles

4. The sodium-potassium pump
   a. is a passive transport process
   b. requires ATP
   c. moves sodium into the cell
   d. moves ions down their gradients

5. When a white blood cell of the immune system is engulfing a bacterium, which process is occurring?
   a. exocytosis
   b. cytolysis
   c. pinocytosis
   d. phagocytosis

EXERCISE
4·3

**Short Answer.** *Write brief responses to the following.*

1. Summarize the cell theory.

_____

_____

_____

_____

_____

_____

2. How are eukaryotic cells distinguishable from prokaryotic ones? What is the relevance of having internal organelles?

_____

_____

_____

_____

_____

_____

3. Laura is lost at sea and only has access to salt water but thinks that she remembers from biology class that she should not drink it. Explain to her what is likely to happen to her blood cells after she has absorbed the salt into her bloodstream.

_____

_____

_____

_____

_____

_____

_____

_____

**EXERCISE**
**4·4**

**Labeling Diagrams.** *Fill in the blanks using the following terms to correctly label the diagram representing a typical eukaryotic plant cell.*

cell wall

central vacuole

chloroplast

nucleus

1._____

2._____

3._____

4._____

**EXERCISE**
**4·5**

**Thinking Thematically.** *For each of the following themes of biology, choose a different concept from this chapter and explain how it provides a useful illustration of that theme.*

1. form facilitates function

_____

_____

_____

_____

_____

_____

2. regulation and feedback

_____

_____

_____

_____

_____

_____

3. continuity and change

_____

_____

_____

_____

_____

_____

## For Further Investigation

Conduct a simple experiment in your kitchen to test the process of osmosis. Soak an unboiled egg in acetic acid (vinegar) for two to three days. Check on the egg periodically, stirring and assisting in the removal of the shell. Eventually, the shell will have completely dissolved. Carefully rinse the egg under a very gentle stream of water and then place it in a cup of syrup overnight. The next day, observe the appearance of your egg. Carefully rinse again, trying to remove as much syrup as possible. Then place the egg in a cup of water overnight (distilled is best, but any old water will do) and again observe your egg. Use the egg as a model for a cell conducting osmosis, applying organelle and transport vocabulary as appropriate.

# Powering Life Processes

## Metabolism

It should now be clear that the cell is the basic unit of structure and function for all living things. What really keeps each cell going is its metabolism—the sum of all the chemical reactions occurring within a living unit and one of the defining characteristics of life. When you hear the term *metabolism*, you may think of the human diet and the different ways people seem to process food. This is a very narrow view of the term. Metabolism involves not only that notion of **catabolism**, the digestion of large molecules into simpler subunits, but it also includes the opposite process of **anabolism**, the building up of larger molecules from smaller ones.

We might consider the metabolism of a cell by examining the way that the sugar glucose is broken down into that all-important cellular energy molecule, ATP. This ATP then provides the energy input for all of the endergonic (energy-requiring) reactions needed to sustain life for each cell. In order to be as energy-efficient as possible, cells often will coordinate an exergonic (energy-releasing) reaction with an endergonic one in a process called **coupling**. Another significant example of cellular metabolism is the photosynthetic process of glucose being produced from just water, carbon dioxide, and energy from sunlight.

## ATP: Currency of Cell Energy

Often when our cells break down macromolecules, they are doing so to release beneficial energy that is stored within the bonds of the molecule. As discussed in Chapter 4, cells are primarily interested in energy in the form of ATP. The ATP molecule is composed of three phosphate groups attached to an organic adenosine molecule (see Figure 5.1). When the terminal phosphate is broken from the chain, the energy held in that bond is released and can be directed to perform useful work for the cell.

Not only is ATP an efficient energy source, it is efficient at what it does. Each time it is used, it is really just broken into two smaller parts, **ADP (adenosine**

**Figure 5.1** An ATP Molecule

**diphosphate**) and $P_i$ (**inorganic phosphate**). If the cell can somehow couple these two pieces with some energy being given off spontaneously by other cellular metabolic reactions, then the two pieces can effectively be put back together again, and ATP can be re-formed and reused for additional cellular work.

# Regulation of Metabolism with Enzymes

Recall that enzymes are biological catalysts composed of protein. Enzymes are structured in such a way that they are specific to only one type of substance, or **substrate**, and can thus catalyze only one type of reaction. This is because each enzyme contains an **active site**, a region that recognizes and binds to the substrate. Although still often referred to as a *lock-and-key interaction*, more recent evidence suggests that it is actually an **induced fit** (see Figure 5.2).

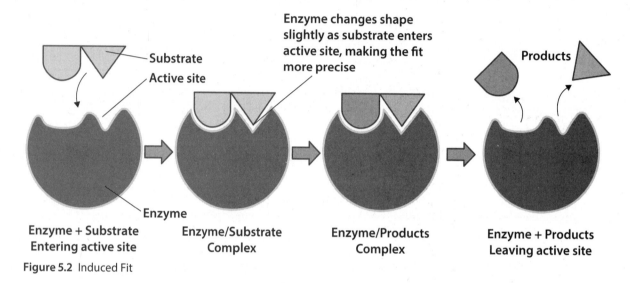

**Figure 5.2** Induced Fit

That is, there is a slight conformational change that occurs when the enzyme's active site and its substrate approach each other. Once binding occurs, the composite structure is called the *enzyme-substrate complex*. Here, the enzyme positions the substrate such that it is energetically favorable to form the product(s) of the reaction. The enzyme thus acts by lowering the **activation energy** necessary to complete the reaction, so the products are reached more quickly (see Figure 5.3). Enzymes are then useful for regulating metabolic reactions. If an enzyme is present, its

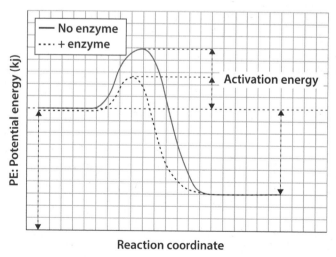

**Figure 5.3** Enzyme Reactions

Create a caption for Figure 5.3 using the terms *enzyme, substrate, active site,* and *activation energy* to compare a catalyzed reaction to an uncatalyzed one.

chemical reaction will occur and desired products will be made. When sufficient product is accumulated, it often provides feedback by binding to the enzyme and slowing or stopping its catalytic action.

# Photosynthesis: Making Sugar with Light

Plants and other **autotrophs** have the ability to make their own organic food in the form of simple sugars in a chemical process called **photosynthesis**. By contrast, **heterotrophs** like animals and fungi must eat other organisms to obtain important organic compounds. Plants comprise an ecologically significant group known as the carbon-fixers, the only organisms that can take the inorganic carbon dioxide ($CO_2$) waste product from other organisms and recycle it into organic glucose, thereby making it useable to others again. They do so within the chloroplasts, specialized organelles that can capture sunlight energy with their brightly colored **pigment** molecules and redirect it into energy shuttles like ATP. This energy is then used to combine that inorganic $CO_2$ with other organic molecules to make glucose, a monosaccharide. Here is the sum reaction representing the entire photosynthetic biochemical pathway:

$$\text{carbon dioxide} + \text{water} \xrightarrow{\text{light}} \text{glucose} + \text{oxygen}$$

Photosynthesis happens in two phases, based on two very different energy transfer processes. The first phase of photosynthesis is known as the **light-dependent reactions**, during which certain wavelengths of sunlight energy are captured by pigments in the chloroplast. True plants are characterized as possessing the green pigment **chlorophyll *a*** as the **main pigment** while also utilizing related **chlorophyll *b*** and other yellow-orange pigments called **carotenoids** as **accessory pigments**. The accessory pigments complement the action of chlorophyll *a* by capturing different wavelengths of light and thus maximizing photosynthetic efficiency. Once pigments have captured some light energy, this energy is used to move electrons from molecule to molecule through a series of specialized reactions called the **electron transport chain**. The final reaction in the chain involves the production of a high-energy molecule called **NADPH**.

While all of that is happening along the folded internal membranes of the chloroplast called **thylakoids**, inside the pockets of these folds the important water molecule water is breaking apart (refer to Figure 5.4). Water generates a constant supply of electrons to replace those that are involved in the reactions that produce NADPH, and also produces hydrogen ions ($H^+$) and oxygen gas ($O_2$). The oxygen is primarily considered a **by-product**, as some is retained for the plant's

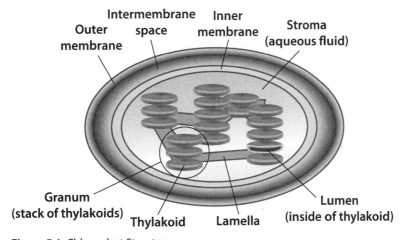

**Figure 5.4** Chloroplast Structure

own cellular respiration needs to break down the glucose it produced, but much of the oxygen is released into the atmosphere and made available to other organisms, like us.

The hydrogen ions are not a by-product, however, as they are directly used to recycle ATP in a process called **chemiosmosis** (see Figure 5.5). As water continues to be broken down and the $H^+$ ions build up in the thylakoid compartment, a concentration gradient is established. The $H^+$ ions cannot simply diffuse through the thylakoid membrane, as they are too charged to make it through the hydrophobic interior. Embedded within that same membrane is a molecule called **ATP synthase**, an enzyme that catalyzes the synthesis of ATP from its component parts (ADP and $P_i$) but also is an $H^+$ ion channel. As a hydrogen ion passively moves through ATP synthase, a conformational change takes place in ATP synthase, resulting in the production of ATP.

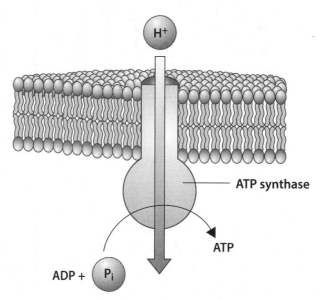

**Figure 5.5** Chemiosmosis

Following the light-dependent reactions are the **light-independent reactions**. These reactions rely on the high-energy products from the light-dependent reactions, namely ATP and NADPH. Recall that the energy to build these molecules ultimately came from the sun. ATP and NADPH act as energy shuttles and are used in the second phase of photosynthesis to build the organic monosaccharide glucose. The set of reactions known as the **Calvin cycle** orchestrates this process, utilizing the inorganic carbon dioxide ($CO_2$) that diffuses in from the atmosphere through tiny pores in plant leaves called **stomata**. The $CO_2$ is combined with other organic compounds, some atoms are rearranged using the energy from ATP and NADPH, and eventually a glucose molecule results. These light-independent reactions are taking place just outside the thylakoid membranes, right where the ATP and NADPH were first produced, in the space inside the chloroplast known as the **stroma**.

# Cellular Respiration: Turning Sugar into ATP

Like photosynthesis, **cellular respiration** is a biochemical pathway utilized by most organisms as part of their metabolism. Instead of building up food energy, however, this process describes the breakdown of that glucose and the extraction of ATP for cellular energy. Cellular respiration is an **aerobic** process, requiring oxygen for completion. For organisms that are **anaerobic** and to whom oxygen is toxic, or for use when some cells within an aerobic multicellular organism face oxygen debt, the less efficient process called **fermentation** takes place in the cytosol. Anaerobic fermentation starts with the process called **glycolysis**, just as aerobic cellular respiration, but takes a different pathway from there in which relatively little overall energy is extracted.

During glycolysis, a molecule of glucose is slowly and systematically broken down and its stored energy extracted. The final chemical product of that breakdown is **pyruvic acid**, and the energy released is stored in two more useful forms: ATP and **nicotinamide adenine dinucleotide (NADH).** While ATP functions as ready energy for all cells in countless ways, NADH acts as a more specific high-energy intermediate that eventually contributes to the production of additional ATP molecules in the final step of cellular respiration.

Pyruvic acid can be involved from there in one of two pathways, depending on the type of organism involved. Organisms like plants, many bacteria, and yeasts undergo **alcoholic fermentation**, a process of converting pyruvic acid into the less toxic **ethanol**. Carbon dioxide gas is also released as a by-product. Other organisms, like mammals with muscle cells in oxygen debt, instead carry out **lactic acid fermentation**. Here, pyruvic acid is converted into **lactic acid**, a substance that can be absorbed by the bloodstream. No excess $CO_2$ is released in this form of fermentation.

When the environment is aerobic and oxygen is available in the cytosol after glycolysis, the fermentation step is avoided altogether and the more energetically favorable process of cellular respiration instead occurs. The pyruvic acid produced at the end of glycolysis moves from the cytosol into a mitochondrion where the remainder of the pathway takes place. First, the pyruvic acid is broken down into an intermediate substance, extracting more NADH and releasing some $CO_2$. The remaining stored energy in this intermediate is finally extracted during a process known as the **citric acid cycle** (or Krebs cycle). During the citric acid cycle, additional molecules of ATP and NADH are extracted and a new energy-intermediate called **flavin adenine dinucleotide (FADH$_2$)** is also generated.

All of this has thus far been taking place in the interior space of the mitochondrion, called the **matrix** (see Figure 5.6). Once all of the NADH and FADH$_2$ have been collected, these energy shuttles participate in a final energy transfer process that includes an electron transport chain and chemiosmosis. Analogous to the location of like processes in the chloroplast, these reactions occur between molecules embedded within the inner membrane of the mitochondrion. Large folds in the membrane called **cristae** greatly increase the surface area of the membrane and thus increase the energy efficiency of the organelle.

Specifically, the NADH and FADH$_2$ are broken down to kick-start the process, as the released energy is used to drive the electron-exchange reactions of this electron transport chain. At the end of the chain, $O_2$ is present, and a reaction generating the by-product water occurs. At the

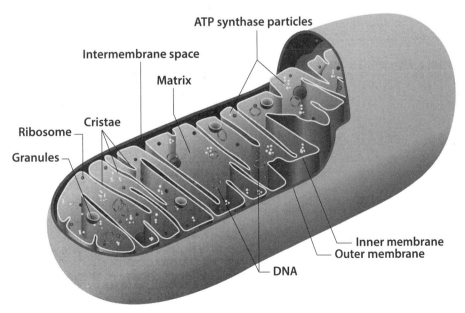

**Figure 5.6** Mitochondrion

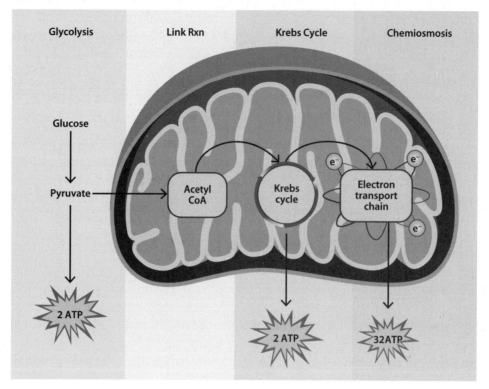

Figure 5.7 A Summary of Glycolysis and Respiration

same time, the breakdown of NADH and FADH$_2$ generates a hydrogen ion gradient. Exactly as in photosynthesis, this H$^+$ gradient drives the process of chemiosmosis. The final result is a high yield of ATP, approximately 36 to 38 molecules per glucose invested, now useful for conducting cellular work (see Figure 5.7).

EXERCISE
5·1

**Vocabulary Building.** *Explain the relationship between the following pairs of vocabulary terms.*

1. enzyme, active site

_____

_____

2. pigment, light reactions

_____

_____

3. mitochondrial matrix, chemiosmosis

_____

_____

**Multiple Choice.** *Select the best response from the options provided to answer each question or to complete each statement.*

1. All of the following are products of the light reactions *except*
   a. oxygen
   b. carbon dioxide
   c. ATP
   d. NADPH

2. Carbon fixation is a special characteristic possessed by
   a. anaerobes
   b. decomposers
   c. autotrophs
   d. heterotrophs

3. Growing the proteins that make up the skin and hair of mammals involves
   a. anabolism
   b. catabolism
   c. dehydration synthesis
   d. both a and c

4. The site of the initial breakdown of glucose that takes place during glycolysis is the
   a. cytosol
   b. cristae
   c. thylakoid
   d. mitochondrial matrix

5. The desired end product of cellular respiration is
   a. NADH
   b. ATP
   c. NADPH
   d. all of the above

**EXERCISE**

**5·3**

**Short Answer.** *Compose concise written responses to the following.*

1. How are enzymes able to function as biological catalysts? Explain the way that energy is involved in the process.

   _____

   _____

   _____

   _____

   _____

2. How are the organelles, the chloroplast and the mitochondrion, well suited for the process of energy transfer?

   _____

   _____

   _____

_____

_____

3. When doing an endurance activity for an extended period of time, some of a person's muscle cells might find themselves in a state of oxygen debt. Cells that have access to oxygen will use it for cellular respiration, but others lacking it will switch to lactic acid fermentation. How does this continue to benefit the body? What will happen if this continues unchecked?

_____

_____

_____

_____

_____

_____

**EXERCISE**

**5·4**

**Interpreting Diagrams.** *Examine the following graph demonstrating the progress of the same reaction with and without an enzyme. Use the information in the graph to answer the questions that follow.*

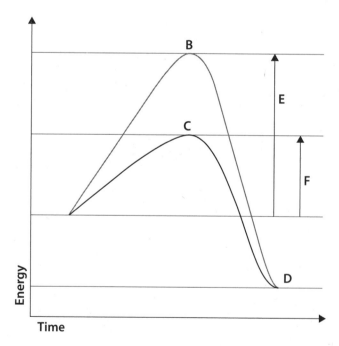

1. Which letter represents the uncatalyzed reaction? _____

2. Which letter represents the catalyzed reaction? _____

3. Which letter represents the products of the reaction? _____

4. Which letter represents the activation energy of the catalyzed reaction? _____

EXERCISE
**5·5**

**Thinking Thematically.** *For each of the following themes of biology, choose a different concept from this chapter and explain how it provides a useful illustration of that theme.*

1. form facilitates function

_____

_____

_____

_____

_____

_____

2. energy and organization

_____

_____

_____

_____

_____

_____

_____

3. natural interdependence

_____

_____

_____

_____

_____

_____

_____

## For Further Investigation

Most bread contains the fungal microorganism yeast. Its action during the baking process contributes to the characteristic rising of the dough. What part of the alcoholic fermentation process is actually responsible? Why can bread dough fall if disrupted before it has completed baking? Some quick Internet research can help confirm your ideas.

# Generations of Life

## Mitosis and Meiosis

Recall that, while DNA is always a polymer of nucleotides, the larger form that the DNA takes can vary. **Chromosomes** were so named because this condensed, highly organized form of DNA is more clearly visible under the microscope when stained. (Imagine thread off or on a spool; it may be difficult to easily see the thread when removed from the spool and stretched thin, but when wrapped completely around a spool, the collective thread is much more easily observed.) Most of a cell's life span, however, is not spent with its DNA in chromosome form. Instead, DNA is usually in **chromatin** form, unraveled and much less apparent under the scope. The reason for these morphs of DNA has to do with the needs and the functions of the cell in the moment. When the cell needs to access its genetic code for copying or for expression of traits, then DNA must be in chromatin form. When the cell undergoes division, however, it must separate the two copies of DNA and allocate them to its offspring cells. Cell division necessitates a more organized form of DNA, thus chromosomes form for the time being.

Chromatin is actually already more highly organized than just the molecular double helix (described in detail in Chapter 7). That double helix wraps around a spherical protein called a histone; this acts to shorten the nearly six feet of DNA present in the nucleus of each human cell (refer to Figure 6.1). In order to organize into a chromosome, the histones stack up upon themselves, and that structure twists and coils, again making its size more compact. Eventually, the DNA has become completely packed into the recognizable chromosome.

## Chromosome Structure

Eukaryotic chromosomes are composed of two identical DNA strands called chromatids (see Figure 6.2). One **chromatid** is made up of the original DNA

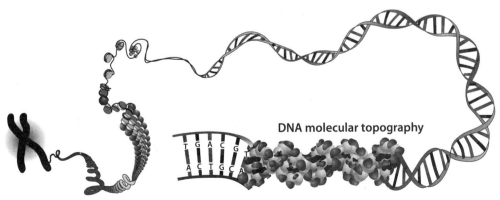

DNA molecular topography

**Figure 6.1** Forms of DNA

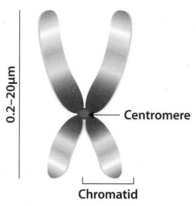

**Figure 6.2** Chromosome Structure

strand possessed by the cell, and the other was made during the S phase of the cell cycle, discussed later in this chapter. When in chromosome form, each identical strand makes up one half of the chromosome and is referred to as a **sister chromatid** to the other. The region within the chromosome where the sister chromatids attach is called the **centromere**.

If you were able to look inside the nucleus of a typical human cell, you would find forty-six total chromosomes. The vast majority of these chromosomes are called autosomes, while only two are considered sex chromosomes. While **autosomes** carry the instructions for most of our genetic traits, the **sex chromosomes** are specifically involved in the determination of biological sex. The two types of sex chromosomes are commonly referred to as the **X chromosome** and the **Y chromosome**.

In humans and all other sexually reproducing organisms, the chromosomes exist in pairs in all body cells. These **homologous pairs** occur because each organism receives half of its DNA from each parent during reproduction. The only pair of chromosomes that may not be truly homologous are the sex chromosomes, as seen in human males with the characteristic XY pattern. All of the autosomes are homologous, having the same size and shape and carrying information for the same types of genes, but the specific instructions may vary depending on the genetics of the parents. Cells that possess two sets of DNA arranged into homologous pairs of chromosomes are referred to as **diploid** cells. When a sexual organism has to make reproductive cells, however, it must split the sets apart into separate **haploid** cells, cells that only have one set of DNA.

# The Cell Cycle

The set of sequential events that make up the life span of an individual cell is called the **cell cycle** (see Figure 6.3). A new cell that has just been created begins its life during **interphase**, a time of metabolism, growth, and expression of genetic traits. Throughout interphase, the DNA is found in chromatin form so that the genetic code is more easily accessible than when wrapped up into a chromosome.

The first portion of interphase called $G_1$ **phase** (gap 1) is characterized by an increase in cell size and an accumulation of ATP. Once this cell achieves an appropriate size, has enough energy, and gets the appropriate signaling from its immediate environment, it will enter the **S phase** (synthesis). The synthesis that takes place here is actually the replication of DNA, an energy-requiring process that will be discussed in detail in Chapter 7. The extra copy of DNA that results prepares the cell for the eventual production of offspring cells. When the S phase is complete, the cell then enters the $G_2$ **phase**. A typical human cell would be considered **diploid** because it has homologous pairs of each type of chromosome present in the nucleus, one inherited from each parent. A human reproductive cell, however, needs to be **haploid** and only contain one set of each chromosome type. That is because one reproductive cell is fertilized by another and their two haploid nuclei will fuse to create a new diploid cell (and organism).

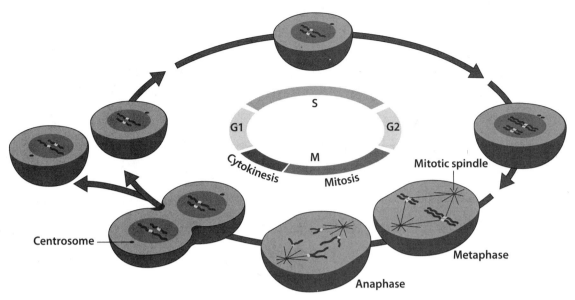

**Figure 6.3** Cell Cycle

# Mitosis: Growing New Body Cells

**Mitosis** specifically involves the carefully orchestrated division of a cell's nucleus and separation of the two copies of DNA that it contains. (This nuclear division is usually–but not always–complemented with a separation of the remainder of the cytoplasm and a fission into two new offspring cells, discussed in detail later in this section.) Mitosis is described in phases based on the state of the cell's DNA at any given moment during its life cycle, which can be observed fairly easily under a light microscope (see Figure 6.4). **Prophase** represents the first phase, during which the cell prepares its DNA and nucleus for division. Recall that the DNA remains in chromatin form from interphase, so immediately the DNA begins the process of condensing itself into a highly organized chromosome. At the same time, the nuclear envelope starts to break down, temporarily allowing the DNA to move more freely around the cell. New structures called **spindle fibers** start to take form, which will eventually provide a framework to direct that movement of DNA. Small centrioles also become apparent and start to migrate away from each other and toward opposite poles of the cell.

Once prophase is complete, the events of metaphase occur. During **metaphase**, the spindle fibers, now attached from the centrioles to each chromosome at its centromere, pull each chromosome toward the equator of the cell, called the **metaphase plate**. Once aligned with their centromeres along the plate, the cell begins **anaphase**. The spindle fibers, still attached, now pull each chromosome apart. One chromatid is pulled toward one pole, while its sister chromatid moves toward the other pole. Once the DNA reaches its respective pole, the cell now enters telophase, the final stage of mitosis.

| Prophase | Metaphase | Anaphase | Telophase | Cytokinesis |

**Figure 6.4** Mitosis and Cytokinesis

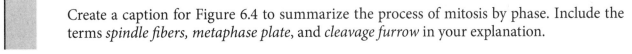

Create a caption for Figure 6.4 to summarize the process of mitosis by phase. Include the terms *spindle fibers*, *metaphase plate*, and *cleavage furrow* in your explanation.

During **telophase**, the cell prepares itself for the next interphase, so many of the events of prophase reverse themselves. Specifically, the DNA unravels back into chromatin, the nuclear envelope and nucleoli re-form, and the centrioles and apparatus of the spindle break down.

At the same time that telophase is taking place and the two new nuclei are becoming apparent, the rest of the cell might start to divide. This process, called **cytokinesis**, involves the splitting of the cytoplasm and the creation of separate cell membranes to establish two individual offspring cells. Animal cells, lacking any external cell wall, carry out cytokinesis through the formation of a **cleavage furrow**. When the individual cell membranes are forming, it appears as if the large parent cell is being pinched down the middle until the cell is split into two separate offspring cells.

The cell wall in plants necessitates a different approach to cytokinesis. A plant cell uses vesicles to deposit cellulose down the midline of the large parent cell, establishing the **cell plate**. Eventually, the new cell plate will merge with existing cell walls, and cytokinesis of the plant cell will be complete.

# Meiosis: Creating Sex Cells

While mitosis is useful for creating new diploid body cells, **meiosis** is required for the production of haploid sex cells, or **gametes** (see Figure 6.5). In order to accomplish the creation of a cell with only one set of DNA, meiosis must involve two sets of nuclear divisions in sequence. The first division, called **meiosis I**, is referred to as the *reduction division* because it successfully reduces the chromosome number in the two resultant nuclei from diploid to haploid. The new haploid offspring cells each still have an extra copy of that DNA, however, so **meiosis II** follows. This nuclear division is described as the mitotic division of haploid cells, as each offspring cell from meiosis I undergoes a series of events just like mitosis except that the initial cells are haploid instead of diploid.

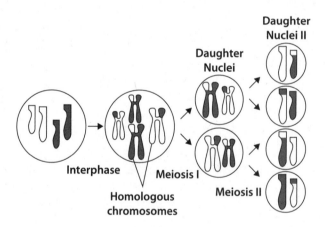

**Figure 6.5** Meiosis Overview

The events of meiosis I begin with **prophase I**. Although bearing many similarities to prophase of mitosis, there are some key points of distinction. First, after the chromosomes form, they arrange themselves into homologous pairs in a process called **synapsis**. This pairing up of homologues consists of four total chromatids, thus it is also referred to as a **tetrad**. Once in tetrad formation, an important event called crossing-over can occur. During **crossing-over**, a segment of a chromatid from one chromosome breaks off and changes places with the homologous piece of a nonsister chromatid from the other chromosome. This results in the shuffling of genetic information that is packaged together on a chromosome before the creation of gametes, an important component of **genetic recombination**. The significance of this is that an individual's gamete will have chromosomes that contain some mix of maternal and paternal chromosomes, newly arranged from the way they were inherited.

After prophase I, the chromosomes are lined up on the equator during **metaphase I**, but they are arranged in homologous pairs. The second type of genetic recombination, **independent assortment**, is set up here as the pairs of chromosomes align themselves randomly with respect to the equator. That is, one set of chromosomes might involve the maternal chromosome aligning on the left and the paternal on the right, but for the pair of chromosomes just below, the opposite situation could occur. Again, this serves to increase genetic variation, an important contribution to the ability of organisms to better adapt and evolve.

During **anaphase I** that follows, the homologous pairs are pulled apart. One entire chromosome is thus moved toward the centrioles at one pole, and the other chromosome of the pair is pulled to the other pole. When **telophase I** is complete, each nucleus has only one set of the original parental chromosomes, and is thus haploid. Each chromosome is still composed of two chromatids, however, thus containing two identical sets of the same haploid DNA.

Meiosis II begins from there, as each haploid cell created during the first division now proceeds through a mitosis-like division. The duplicated chromosomes still present in each haploid cell are lined up on the equator, pulled apart, and placed into one of two new offspring cells created. The final product of meiosis is four genetically unique haploid cells with unduplicated DNA.

While the events of meiosis are responsible for the production of all gametes, there are special pathways that result in the unique features of sperm and eggs. During **spermatogenesis**, equal allocation of cytoplasm during all cytokinesis results in the production of four haploid, equally sized but genetically variable sperm cells. Differentiation of each cell results in the familiar flagellum (tail) and reduced cytoplasm.

When female gametes are instead being created during **oogenesis**, the cytoplasm is not split evenly. In fact, one cell repetitively receives the vast majority of the cytoplasm, while the other is short-changed. In the end, this results in only one functional haploid egg but also produces three other haploid cells called **polar bodies**. The polar bodies are not functional gametes and will degenerate after production.

**EXERCISE 6·1**

**Vocabulary Building.** *Provide a definition for each of the following vocabulary terms. When possible, identify any roots in the term and use them to help create the definition.*

1. autosome

_____

_____

2. diploid

_____

_____

3. interphase

_____

_____

4. cytokinesis

_____

_____

5. oogenesis

_____

_____

_____

EXERCISE
**6·2**

**Multiple Choice.** *Select the best response from the options provided to answer each question or to complete each statement.*

1. All of the following are expected to be associated with DNA during interphase of the cell cycle *except*

   a. histones            c. double helix

   b. chromosome      d. chromatin

2. The offspring cells at the end of mitosis are

   a. diploid            c. genetically unique

   b. gametes        d. both a and c

3. The phase of the cell cycle during which the quantity of DNA doubles is

   a. mitosis           c. synthesis

   b. gap 1           d. gap 2

4. During which of the following phases would you be expected to find enough DNA to comprise homologous pairs of chromosomes?

   a. metaphase       c. metaphase II

   b. metaphase I      d. both a and b

5. Which of the following is *not* an example of a haploid cell?

   a. gamete          c. skin cell

   b. sperm          d. bacterium

EXERCISE
**6·3**

**Short Answer.** *Write brief responses to the following.*

1. Explain the difference in the cytokinesis that follows mitosis of plant and animal cells.

_____

_____

_____

_____

_____

_____

2. Mitosis is not always followed by cytokinesis (i.e., complete cell division does not always occur). Explain what would result when this happens.

_____

_____

_____

_____

_____

_____

3. What is the significance of the focus on the quality of just one of the four cells that result from oogenesis? Why is there a different approach during spermatogenesis?

_____

_____

_____

_____

_____

_____

_____

**EXERCISE**
**6·4**

**Labeling Diagrams.** *Fill in the blanks using the terms from the following list to correctly label the diagram representing cells in various phases of mitosis.*

anaphase

metaphase

prophase

telophase

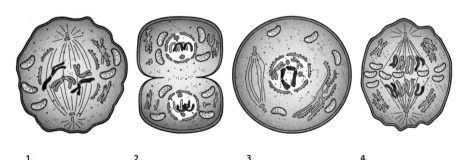

1. _____   2. _____   3. _____   4. _____

**Thinking Thematically.** *For each of the following themes of biology, choose a different concept from this chapter and explain how it provides a useful illustration of that theme.*

1. regulation and feedback

_____

_____

_____

_____

_____

2. energy and organization

_____

_____

_____

_____

_____

3. continuity and change

_____

_____

_____

_____

_____

## For Further Investigation

The next time you suffer a minor cut or scrape, try to imagine the cell repair process that immediately starts to take place. Does mitosis seem to happen at different rates depending on the location of the cut on the body or the condition the wound is placed in? What might be affecting any difference observed? Consider why a scar may form. Hypothesize as to why adults tend to scar less frequently than children.

# GENETICS

# DNA and the Central Dogma of Molecular Biology

One of the most significant discoveries ever made in the field of biology involved unraveling the exact structure of the DNA molecule. **DNA** (recall its full name is deoxyribonucleic acid) is so significant because it carries the full set of genetic instructions necessary for all life on Earth. While the four basic monomers present in DNA are the same from the simplest bacteria to a complex mammal, the specific sequence into which those monomers are arranged to compose a **gene**, and the total number, arrangement, and type of genes present in an organism's **genome** varies widely between species. In the end, the secret to the world's vast biodiversity throughout various ecosystems is rooted in the genetic diversity present among genomes.

The discovery of DNA's structure is often still credited solely to James Waton and Francis Crick; however, a crucial piece of primary data came from contemporaneous scientist Rosalind Franklin. She worked tirelessly to understand the structure of DNA through a technique called X-ray crystallography, which revealed some internal features of the molecule that Watson and Crick then incorporated into their models.

## Nucleotides: The Building Blocks of DNA

Recall that DNA is composed on the simplest level of compound monomers called nucleotides and that each nucleotide consists of a sugar (deoxyribose), a phosphate group, and a nitrogenous base. The nitrogenous base varies between nucleotides and provides the distinct character for each, so nucleotides are named by the type of nitrogenous base present. Two of the nitrogenous bases, adenine (A) and guanine (G), are composed of two fused, carbon-based rings and are referred to as **purines** (see Figure 7.1). The other remaining bases, cytosine (C) and thymine (T), are composed of a single carbon-based ring and are called **pyrimidines**.

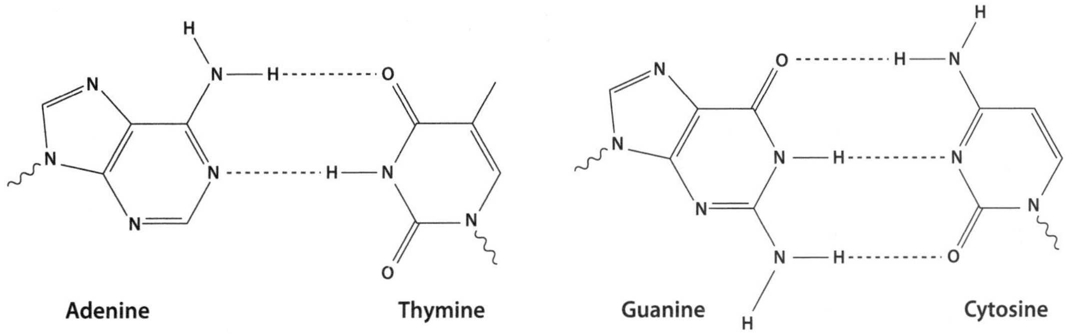

Adenine     Thymine     Guanine     Cytosine

**Figure 7.1** Complementary Base Pairing

A nucleotide is covalently bonded to adjacent nucleotides in such a way as to create a sugar-phosphate backbone with exposed nitrogenous bases in a single DNA strand (see Figure 7.2). DNA is always arranged in a more complex molecular fashion, however, with two individual strands attaching to each other through an attraction between nitrogenous bases. In order for the nitrogenous bases to have the right molecular attraction, they must be aligned so that they allow for several weak hydrogen bonds to form. Nucleotides that will readily hydrogen-bond to one another

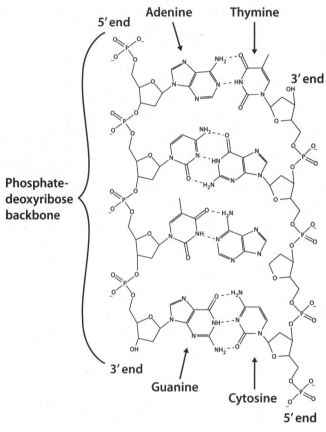

**Figure 7.2a** DNA Double Helix

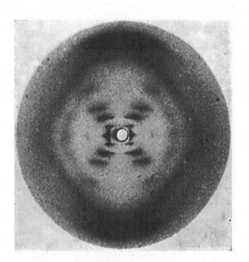

**Figure 7.2b** Photo 51, an x-ray diffraction image of the DNA double helix

https://en.wikipedia.org/wiki/Photo_51#/media/
File:Photo_51_x-ray_diffraction_image.jpg

create a **complementary base pair**. Through experimentation, it was first discovered that the percentage of cytosine and guanine were always equivalent in any DNA molecule, as was the percentage of thymine and adenine. Later, it was determined that cytosine always pairs with guanine (via three hydrogen bonds) and adenine always pairs with thymine (via two hydrogen bonds).

# DNA Replication

The DNA molecule can be thought of like a spiraling ladder, where the sugar and the phosphate portions of the nucleotides align themselves along the sides of the ladder with the sugar positioned at the attachment point for the rungs. The nitrogenous base that is attached to each sugar points toward the inside of the ladder and creates half of each rung. The other half comes from the complementary nucleotide on the opposite strand, which also provides the other side of the ladder via its sugar-phosphate backbone.

The fact that each DNA molecule comprises two complementary strands allows for the molecule to be copied precisely, or **replicated**, for ongoing generations of cells and organisms. During **DNA replication** (Figure 7.3), the double helix is untwisted and split apart at various points with the help of the enzyme **helicase**. As helicase catalyzes the breaking of the hydrogen bonds and the resultant exposure of the nucleotide sequence along either stand, it also creates **replication forks**, sites where DNA replication can begin. Another enzyme, **DNA polymerase**, binds to an area along each single strand. It proceeds to "read" the first nucleotide along the template strand, and then positions the correct complementary nucleotide so that it can hydrogen-bond. The next nucleotide is read, the

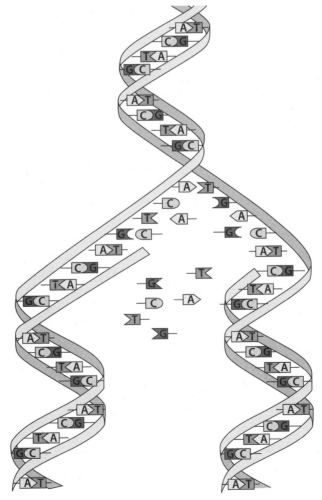

**Figure 7.3** DNA Replication

complementary nucleotide positioned, and then the two new nucleotides in the growing strand are covalently bonded to each other. The process continues until each strand of the original DNA molecule has been completely copied accordingly.

DNA replication is much more complicated when examined in further detail. After helicase unzips the original double helix, the two newly separated strands must be kept from re-hydrogen bonding through the action of single-strand binding proteins. Additionally, because the two strands of nucleotides within one DNA molecule run in opposite directions (i.e., are **antiparallel**), only one strand can be read and its complement built in a very straightforward manner. This is the **leading strand** and its action basically follows in the same direction of helicase, building one new, continuous complementary strand of DNA. The other original DNA strand is called the **lagging strand**, and it runs in the opposite direction from helicase and instead builds a series of segments (called **Okazaki fragments**) that are eventually bonded together through the action of an enzyme called **ligase**.

Why must the two DNA strands be antiparallel and complicate replication so much? As it turns out, the hydrogen bonds that are so essential in complementary base pairing only form correctly to create a stable DNA molecule when its two strands run in opposite directions. This fact actually complicated the work of the researchers working to first understand the structure of DNA in the 1950s, as it was quite unexpected. A now famous image called photo 51, taken in Rosalin Franklin's lab (see Figure 7.2a), finally provided evidence for this antiparallel design.

# Protein Synthesis and the Central Dogma

While DNA is essential in storing and protecting the genetic code, RNA is really the molecule more actively involved in the expression of traits. An RNA nucleotide differs from a DNA nucleotide in a few important ways. First, RNA is composed of the sugar ribose instead of deoxyribose. Second, the nitrogenous base U (uracil), also a pyrimidine, replaces T. Finally, RNA is always single stranded such that at least some of its sequence of nitrogenous bases is exposed.

The **central dogma of molecular biology** is a model that describes the information flow from DNA to RNA to protein during the process called **protein synthesis** (see Figure 7.4). As already described, DNA can synthesize an exact copy of itself through replication, but the sequential processes of transcription and translation are required to produce the proteins that are either directly or indirectly responsible for the expression of our traits. The process of **transcription** involves the reading of a DNA gene and building a complementary set of instructions in the RNA transcript. **Translation** is the subsequent process of reading that transcript and then using those instructions to assemble the specific sequence of amino acids into a protein.

Although always composed of a single strand of RNA nucleotides, the RNA molecule does take on more complex forms depending on the specific role it is playing. There are in fact three primary forms of RNA involved in protein synthesis. The RNA transcript is the least complex, composed of a simple linear strand of nucleotides called **mRNA (messenger RNA)**. The other forms of RNA involved in reading the mRNA during translation are **tRNA (transfer RNA)** and **rRNA (ribosomal RNA)**. The tRNA molecules are described as clover-leaf shaped, for the single strand folds up on itself and creates three loops held together by hydrogen bonds. The rRNA molecule associates with other protein molecules to assemble the overall ribosome structure. In the case of rRNA, the single strand of nucleotides folds up on itself in a more globular fashion.

If the term mRNA is sounding very familiar to you, it may be because some of the vaccines recently created to combat the COVID-19 virus were based on mRNA molecules. This was a new approach that turned out to be very successful indeed and will be discussed more in Chapter 12.

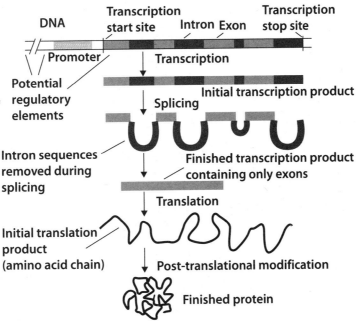

Figure 7.4 Central Dogma in Eukaryotes

# Transcription

The molecular process of transcription takes place wherever the DNA is found. In a prokaryote, this would be the nucleoid region, whereas the nucleus encloses the DNA within a eukaryotic cell and is thus the site of the process. In any case, the DNA gene that needs to be expressed is first located within the genome when the enzyme **RNA polymerase** binds to a specific region called the **promoter**. This triggers the assemblage of some other molecules around the promoter called **transcription factors**, which in turn kick-start the action of RNA polymerase. RNA polymerase forces apart the hydrogen bonds between complementary DNA strands, reads the sequence along one strand, and uses that strand as a template to build a new complementary **mRNA transcript** (see Figure 7.5). This process continues until the RNA polymerase reaches the **termination sequence** at the end of the gene. The transcript loses its affinity for the DNA molecule, which zips back up as hydrogen bonds re-form between complements.

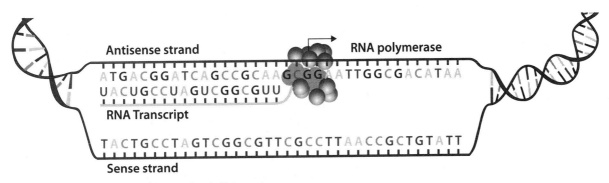

Figure 7.5 Transcription and Processing in Eukaryotes

# Translation

In a prokaryote, the process of translating the mRNA transcript into a specific protein molecule occurs while the transcript is still being built. This is because there is no physical separation of the DNA needed for transcription and the ribosomes necessary for translation. In eukaryotes, however, the processes are kept distinct by the presence of the nucleus. The transcript must exit the nucleus through a nuclear pore before encountering a ribosome either free-floating in the cytoplasm or embedded along the ER.

The ribosome coordinates the action of translation by providing a site for the reading of mRNA and the docking of complementary tRNA molecules (see Figure 7.6). The mRNA molecule is read three nucleotides at a time, a unit called a **codon**. The codon that initiates the growth of a new protein is called the **start codon**. This is necessary because any codon sequence within mRNA is complementary to a specific **anticodon** sequence within a tRNA molecule. The tRNA molecules are responsible for delivering the correct amino acid called for by the codon to the new polypeptide chain being created. To summarize, the code within mRNA is read, and the tRNA molecules work to deliver amino acids to the ribosome where they bond together to elongate the protein until a stop codon is read, terminating the process. The protein itself might be a structural component of the trait itself, like the building of muscle tissue. The protein may instead be functional, as seen in an enzyme that is involved in the production of pigments for eye color.

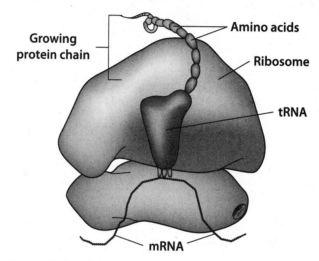

**Figure 7.6** Translation

Create a caption for Figure 7.6 to explain how the three different forms of RNA work together during translation.

**EXERCISE 7·1**

**Vocabulary Building.** *Provide a definition for each of the following vocabulary terms. When possible, identify any roots in the term and use them to help create the definition.*

1. mutation

_____

_____

2. DNA replication

_____

_____

3. RNA polymerase

_____

_____

4. transcription

_____

_____

5. genome

_____

_____

**Multiple Choice.** *Select the best response from the options provided to answer each question or to complete each statement.*

1. Which of the following is *not* involved with DNA replication?
   a. DNA polymerase
   b. helicase
   c. RNA polymerase
   d. free DNA nucleotides

2. Which of the following pairs of a molecular genetics process and its location(s) in a eukaryotic cell is *incorrectly* matched?
   a. translation, cytosol
   b. replication, nucleus
   c. transcription, cytosol
   d. protein synthesis, nucleus and cytosol

3. If the percentage of thymine nucleotides present in a sample of DNA is determined to be 18 percent, which of the following must also be true?
   a. cytosine = 32 percent
   b. adenine = 18 percent
   c. guanine = 18 percent
   d. both a and b

4. Which of the following correctly summarizes the sequence of events that occur during translation?
   a. termination, elongation, initiation
   b. initiation, elongation, termination
   c. elongation, initiation, termination
   d. initiation, termination, elongation

5. The central dogma of molecular biology describes
   a. how genes are expressed
   b. the laws of genetics
   c. the way genes are inherited
   d. how proteins are shaped

**Short Answer.** *Write brief responses to the following.*

1. How do the strategically placed covalent and hydrogen bonds present in DNA facilitate the functioning of the molecule? Be specific for each type of bond.

   _____

   _____

   _____

   _____

   _____

   _____

   _____

2. Use the original DNA sequence that follows to answer the following.

   T – A – C – A – G – G – G – T – A –T – A – A – C – T – G – A – T – C

   a. List the sequence of bases that should result from *replication* of the original DNA strand.

   _____

   b. List the sequence of bases in the mRNA transcript that would be produced by *transcription* of the original DNA strand.

   _____

   c. List the series of tRNA anticodons that are involved in *translation* of the mRNA transcript produced in part b.

   _____

3. Explain why the DNA molecule must be double stranded but RNA (whether mRNA, tRNA, or rRNA) should only consist of a single strand of nucleotides.

   _____

   _____

   _____

   _____

   _____

   _____

**Interpreting Diagrams.** *Examine the following diagram representing the central dogma of molecular biology. Use the information to answer the questions that follow.*

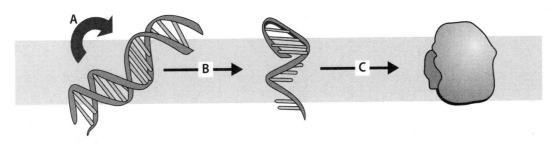

1. Arrow A is representing which process?  _____

2. Arrow B is representing which process?  _____

3. Arrow C is representing which process?  _____

**Thinking Thematically.** *For each of the following themes of biology, choose a different concept from this chapter and explain how it provides a useful illustration of that theme.*

1. form facilitates function

_____

_____

_____

_____

_____

_____

2. energy and organization

_____

_____

_____

_____

_____

_____

_____

3. continuity and change

_____

_____

_____

_____

_____

_____

## For Further Investigation

Investigate the condition of lactose intolerance, an inability to produce the correct protein that digests the milk sugar. Explain the genetic basis of this condition and how it applies to the central dogma concept. Determine whether you are lactose intolerant or lactase persistant. The website biointeractive.org may be particularly helpful in your investigation.

# Mendel and Classical Genetics

Early in the second half of the 1800s, an Austrian monk named Gregor Mendel conducted scientific experiments on pea plants in the isolation of his monastery garden. Curious about the way that pea plants transmitted physical **traits** from one generation to the next, Mendel directed the pollination of thousands of pea plants over many generations in order to uncover the essence of what would later be called the field of **genetics**. Like many great minds, Mendel's work went unrecognized by science until the early twentieth century, and the significance not fully understood until the discovery of the structure of DNA and the molecular genetics revolution that began in the 1950s. Eventually, three laws of **heredity** were established in Mendel's name as a tribute to his lasting contributions to biology.

## Mendel's Methods

Mendel, like most people at the time, understood to some extent that many observable, physical traits were passed from parents to offspring through the process of sexual reproduction and the union of sex cells like sperm and eggs. In pea plants, Mendel observed that traits like flower color, pod shape, and plant height occurred in two discrete forms. For example, flower color was only seen in purple or white form, pod shape as either inflated or constricted, and plant height as only tall or short (see Figure 8.1). Mendel was curious about the predictability of this heredity; why was it that on some occasions two purple-flowered plants would only produce purple-flowered offspring, While in other instances two purple-flowered parents could produce some offspring with purple flowers and others with white flowers?

First, Mendel knew that it was essential to control the trait that each parent plant was able to pass on to offspring. He bred **true-breeding** plants by forcing a plant to **self-pollinate** and then verifying the results after several generations. For example, Mendel took pollen from a tall plant and used it to fertilize its own eggs (this then establishes the parental, or **P generation**). If the resulting offspring in the first generation (this is the first filial, or **F$_1$ generation**) were all tall, then those would be self-pollinated to verify that they can only produce tall offspring in the second generation (second filial, or **F$_2$ generation**). If instead some of the F$_1$ offspring were tall and others short, then that parent plant would be established as not true-breeding and abandoned for research purposes.

Once Mendel had true-breeding pea plants for all of the traits of interest, he then mated another true-breeding plant with the alternate trait. For example, a true-breeding plant for round seeds would be **cross-pollinated** with a true-breeding plant for wrinkled seeds. Mendel found that all of the resulting offspring were always

| Seed | | Flower | Pod | | Stem | |
| Form | Color | Color | Form | Color | Place | Size |
| Round | Yellow | Violet | Full | Green | Axial pods, Flowers along | Long (6–7 ft) |
| Wrinkled | Green | White | Constricted | Yellow | Terminal pods, Flowers top | Short (–1 ft) |
| 1 | 2 | 3 | 4 | 5 | 6 | 7 |

**Figure 8.1** Traits in Pea Plants

round-seeded. Even more interesting to Mendel, however, were the results of the $F_2$ generation. When one of the round-seeded offspring in the $F_1$ generation was crossed with another, they produced some offspring with round seeds and other offspring with wrinkled seeds.

# Predictions and the Punnett Square

When Mendel observed pea plants for their flower color or the height of their stems, he was observing their **phenotypes**, the physical expression of a gene. (It is important to remember that Mendel would have never used the term *gene*, as it hadn't yet been coined, and the structure of DNA was far from understood.) Each phenotype is the result of the combination of **alleles**, or particular forms of the gene, that was inherited from the parents. Collectively, both alleles that an individual possesses for a given trait are described as the **genotype**.

If an individual's genotype is composed of two of the same alleles for a trait, then it is described as **homozygous**. When the like alleles carry information for the dominant trait, then the genotype is more specifically **homozygous dominant**. Likewise, when both alleles are recessive, then the genotype is **homozygous recessive**. If an individual in fact has two different alleles for a given trait, then the genotype is described as **heterozygous** instead.

When the inheritance of one particular trait is under investigation, the cross between two parents heterozygous for that trait is described as **monohybrid**. The cross is traditionally set up in a framework for prediction of offspring called the **Punnett square**. To set up the Punnett square, the genotypes of each parent must be known. One genotype is written at the top of the square with each allele placed in separate columns. The other genotype is written on the left side with each allele placed in separate rows. Then, the cross is performed by matching up the alleles in each column with the allele in each row, and then writing the resultant genotype in the intersecting cell (see Figure 8.2). The four new genotypes represent the four possible combinations of alleles that an offspring of the original parents could themselves possess.

The results of the Punnett square are analyzed and genotype and phenotype ratios are usually described. For example, in Figure 8.2, the Punnett square shows that 1 *PP* genotype, 2 *Pp* genotypes, and 1 *pp* genotypes are possible from the parents. The **genotypic ratio** is then described as 1:2:1. For those described genotypes, three code for the dominant phenotype (1 *PP* and 2 *Pp*), while only one codes for the recessive phenotype (*pp*). The **phenotypic ratio** is thus described as 3:1.

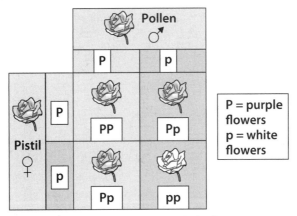

**Figure 8.2** Punnett Square: A Monohybrid Cross

Create a caption for Figure 8.2 by explaining how Mendel's law of segregation is represented in a Punnett square.

## Mendel's Laws of Inheritance

As already mentioned, Mendel is credited with developing what are now known as the laws of inheritance. First, the **law of dominance** holds that the presence of a **dominant allele** completely masks any expression of the **recessive allele** if present (as observed in the heterozygous genotype *Pp* above). The **law of segregation** says that only one of the two alleles that each parent has for a genotype is passed on to offspring at any one time (i.e., through one gamete). This is demonstrated in a Punnett square when the two alleles from one parental genotype are written in separate columns or rows.

According to Mendel's **law of independent assortment**, the expression of one trait does not affect the expression of another, evidenced by the way that the maternal and paternal chromosomes within a homologous pair align along the equator in metaphase I of meiosis. Mendel did not realize it at the time because the traits he was studying happened to all be located on different chromosomes, but the law of independent assortment does not hold true for genes that are **linked**, or housed on the same physical chromosome.

## More Complicated Predictions

When an individual possesses the recessive phenotype, then the genotype is also known to be homozygous recessive (*aa*). When an individual instead has the dominant phenotype, that individual could be either homozygous (*AA*) or heterozygous (*Aa*). In order to determine that dominant individual's genotype, a testcross can be carried out. In a **testcross**, the dominant individual is mated with a homozygous recessive individual (known by their recessive phenotype), and their offspring are analyzed. If the dominant individual is in fact homozygous, then all of the offspring produced from the testcross will have the dominant phenotype. If instead the individual is heterozygous, then the offspring should be a mix of some dominant and some recessive phenotypes.

Often geneticists are not only concerned with the inheritance of one trait but are interested in the way that two traits are inherited together. In that case, a larger Punnett square is utilized to carry out a **dihybrid cross**. The basic principles are the same, but an extra step is required to first establish the combinations of alleles that each parent has for the two different genes under examination. For example, if one parent is heterozygous for both genes (i.e., *Aa* and *Bb*), then that parent could pass along alleles in the combinations of *AB*, *Ab*, *aB*, or *ab*. (Remember that the *A* and *B* alleles are representing different genes, so the law of segregation is not being violated here.)

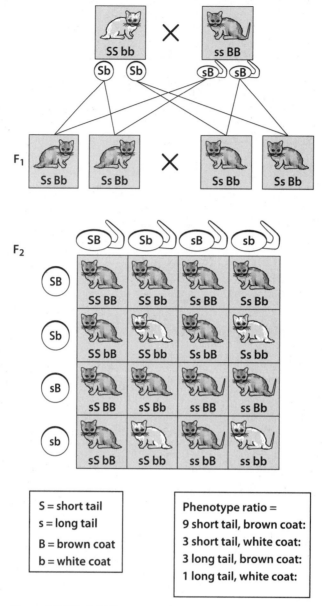

| | SB | Sb | sB | sb |
|---|---|---|---|---|
| **SB** | SS BB | SS Bb | Ss BB | Ss Bb |
| **Sb** | SS bB | SS bb | Ss bB | Ss bb |
| **sB** | sS BB | sS Bb | ss BB | ss Bb |
| **sb** | sS bB | sS bb | ss bB | ss bb |

S = short tail
s = long tail

B = brown coat
b = white coat

Phenotype ratio =
9 short tail, brown coat:
3 short tail, white coat:
3 long tail, brown coat:
1 long tail, white coat:

**Figure 8.3** Dihybrid Cross

The combinations of alleles for the other parent are determined in the same fashion, then arranged in a four-by-four Punnett square (see Figure 8.3).

Because many different genotypic outcomes are possible in a dihybrid situation, often only the phenotypic ratios are calculated. In this case, the phenotypic ratio is 9:3:3:1, signifying that nine of the 16 potential offspring outcomes are dominant for both traits, three are dominant for the first trait and recessive for the second, three are instead recessive for the first and dominant for the second, and only one is recessive for both traits.

## Special Cases of Inheritance

It is probably not surprising to learn that many traits are not inherited according to Mendel's law of dominance. Often, there are two alleles for a trait, but one does not mask the other when they occur together in a heterozygous genotype and a new phenotype instead emerges.

In some cases, a special type of inheritance called **incomplete dominance** specifically occurs. The snapdragon plant exhibits this pattern of inheritance in terms of its flower color.

There are two alleles for flower color in snapdragons, red (*R*) and white (*W*). When two red alleles exist in an individual, then that plant has the homozygous genotype *RR* and exists as a red-flowered phenotype. The same predictable pattern is true for a plant with two white alleles for flower color (genotype = *WW*, phenotype = white). When one red allele and one white allele exist together in an individual as the heterozygous genotype (*RW*), however, the corresponding phenotype is pink.

Although it might at first seem as if the heterozygous pink offspring demonstrates a blending of its homozygous red and white parents, in reality the offspring show alternating red and white pigment cells on such a small scale that the individual cells cannot be distinguished from one another macroscopically. Like individual pixels on a computer screen that blend together into continuous color when the scale increases significantly, the deposition of red pigment next to white simply looks pink to the naked eye.

The coat color of some cows is inherited in another special case called **codominance**. In codominance, both homozygous parental phenotypes are expressed in an individual heterozygous offspring, usually in some sort of macroscopic pattern. A cow with a completely red (rust-colored) coat has the homozygous genotype *RR*, while a cow with a pure white coat has the other homozygous genotype *WW*. If a red cow mates with is a white one, all of their offspring will be heterozygous *RW* and roan, a red-coated cow with white splotches.

Other special cases of inheritance only get more complicated. When **pleiotropy** is observed, as in humans with sickle-cell disease (discussed in more detail in Chapter 9), one gene (in this case, that for hemoglobin), is responsible for the determination of many traits (here, the multiple symptoms experienced by sickle-cell sufferers). Other traits are considered **polygenic**, for the expression of a single phenotype is the result of the interaction of the products of many genes. Any traits describable on a spectrum (human traits like height, eye color, and skin color) are usually polygenic in nature.

In yet another situation called **epistasis**, the expression of one gene actually blocks the expression of another, thus affecting the observed phenotype in interesting ways. An example of epistasis is seen in Labrador retriever coat colors. The presence of the dominant yellow allele in one gene for coat color completely blocks the expression of dark pigment found at another gene, no matter whether the dark pigment were to produce the dominant black or the recessive brown.

**EXERCISE**

**8·1**

**Vocabulary Building.** *Explain the relationship between the following pairs of vocabulary terms.*

1. genetics, heredity

_____

_____

2. genotype, phenotype

_____

_____

3. dominant, recessive

_____

_____

_____

**Multiple Choice.** _Select the best response from the options provided to answer each question or to complete each statement._

1. An individual pea plant possesses an allele for white flowers and an allele for purple flowers. This genotype of this plant for flower color is described as
   a. homozygous dominant
   b. heterozygous
   c. homozygous recessive
   d. codominant

2. The individual pea plant just described in #1 would have what phenotype?
   a. dominant
   b. white
   c. purple
   d. both a and c

3. A human with curly hair has offspring with a person with straight hair. All of their offspring have wavy hair. What type of inheritance pattern is likely being observed?
   a. complete dominance
   b. pleiotropy
   c. incomplete dominance
   d. codominance

4. A plant that is homozygous for axial flowers is bred with a plant that is homozygous for terminal flowers. The offspring should be
   a. 100 percent axial
   b. 100 percent terminal
   c. 50 percent axial, 50 percent terminal
   d. 75 percent axial, 25 percent terminal

5. A mouse has two dominant black alleles for coat color but also possesses the genes to produce the hairless condition. Which special type of genetic inheritance is observed in this case?
   a. pleiotropy
   b. epistasis
   c. polygenic
   d. codominance

**Short Answer.** _Write brief responses to the following._

1. Explain how the process of meiosis and the creation of gametes demonstrate Mendel's laws of segregation and independent assortment.

_____

_____

_____

_____

_____

_____

2. Can a roan cow be used in a testcross? Explain.

_____

_____

_____

_____

_____

_____

3. A pink snapdragon is cross-pollinated with a white snapdragon. If they produce 20 offspring, determine the possible phenotypes and the number of individual offspring that would be predicted to have each phenotype.

_____

_____

_____

_____

_____

_____

**Interpreting Diagrams.** *Complete the following Punnett square using the information provided. Then answer the questions that follow.*

*A pea plant heterozygous for axial flowers is cross-pollinated with a plant with terminal flowers.*

1. What is the genotype of the parent with terminal flowers?

   _____

2. What is the genotype ratio of the offspring in the $F_1$?

   _____

3. What is the phenotype ratio of the offspring in the $F_1$?

   _____

4. If the original parents produced 100 offspring, approximately how many would have axial flowers?

   _____

EXERCISE

8·5

**Thinking Thematically.** *For each of the following themes of biology, choose a different concept from this chapter and explain how it provides a useful illustration of that theme.*

1. science methodologies and applications to society

   _____

   _____

   _____

   _____

   _____

   _____

   _____

2. continuity and change

   _____

   _____

   _____

   _____

   _____

   _____

   _____

3. natural interdependence

   _____

   _____

   _____

_____

_____

_____

## For Further Investigation

Conduct a simple genetic analysis of yourself for a few genetic traits that are easily character-ized. First, carefully examine the middle digit of your fourth finger (the "ring finger") for the presence of any tiny hairs. If you have the dominant "mid-digital hair" trait, then you know you have at least one dominant allele and your genotype can be described as $M\_$. If instead you lack any hair, you have the recessive phenotype, and your genotype must be *mm*. Sec-ond, pull your hairline back and look in the mirror. Does your hairline create a distinct point near the center called a widow's peak? If so, you possess the dominant trait. In any case, what is your genotype? (If you have the dominant phenotype and have access to your own parents and/or offspring, then you may be able to determine your exact genotype with some careful analysis.)

# Modern Genetics and Biotechnologies

·9·

In the early 1900s, geneticist Thomas Hunt Morgan unknowingly led an effort that modernized the field of genetics first launched by Mendel. He and his lab focused on the genetics of the common fruit fly, *Drosophila melanogaster*. Through their careful work, Morgan's lab uncovered two very important discoveries in the field: first, the presence of certain chromosomes determine biological sex, and second, some traits are related to the biological sex of the organism. Although Morgan was focused on the genetics of a tiny insect, much of what he learned continues to be applicable to the genetics of other organisms, including humans. The field of human genetics has continued to expand, as new modes of detection are employed and novel mechanisms for manipulating nucleic acids and proteins turn science fiction into reality. The field of forensic genetics has also found a home in pop culture, with crime-solving television shows, films, and novels accessible to all. Most notably, forensic geneticists have recently helped solve several major cold cases, including the long-awaited identification of the Golden State Killer.

## Sex Chromosomes and Sex-Linked Traits

When Morgan observed the chromosomes of *Drosophila* under the microscope, he was able to verify that the chromosomes occurred in pairs (i.e., the cells were diploid). While the pairs seemed perfectly matched in females, one chromosome was noticeably shorter than its assumed partner chromosome in males. What Morgan was observing were the sex chromosomes, those present in many animals that carry genes that help determine the biological sex of the organism. Like in mammals, it was determined that possessing two homologous X chromosomes makes for the female condition, while possessing one X chromosome along with a much smaller Y chromosome creates the male sex.

The results for some of the *Drosophila* crosses performed in Morgan's lab posed some interesting questions to the researchers. Although several eye colors are known to exist in fruit flies, the dominant and **wild-type** (common naturally occurring) phenotype is red. The researchers were confused as to why, when a red-eyed female fruit fly was mated with a red-eyed male, all of the female flies produced had the expected red eyes, but half of the males surprisingly possessed white eyes. What the scientists were witnessing was the effect of **sex-linked genes**, traits whose instructions are carried on a sex chromosome and thereby linked to the individual's biological sex.

Because the gene for eye color in fruit flies is found on the X chromosome, females (*XX*) have two alleles for eye color, while males (*XY*) only have one, a condition referred to as **hemizygous**. The allele is thus expressed as a superscript on the X chromosome, while the Y chromosome in males will not include any special notation (because it does not carry homologous information for this gene).

Sex-linked traits are common in humans as well as fruit flies (see Figure 9.1). These include the genetic disorders of red-green color blindness, the inability to distinguish between these two colors, and hemophilia, the inability to form blood clots. Because both of these genes are carried on the X chromosome, males only have one allele for the trait and will produce their phenotype based on that single allele. It is thus more likely for males to demonstrate either of these conditions than females, who gain some protection from the presence of the other homologous X chromosome.

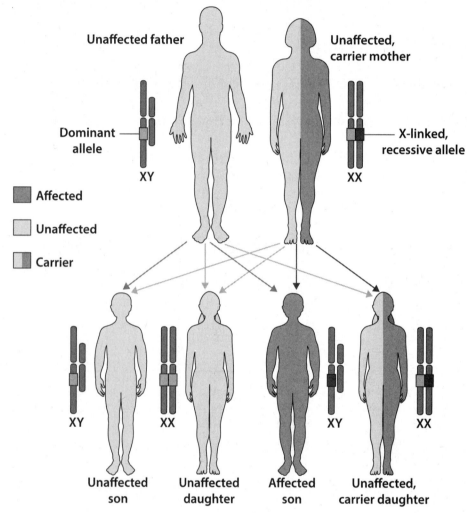

**Figure 9.1** Sex-Linked Cross

Create a caption to Figure 9.1 to explain how a biological son could inherit a sex-linked trait from his maternal line, but not from his paternal side.

# Mutation: Change in the Code

Part of the reason that Morgan was able to uncover so much about the genetics of the fruit fly is that he was able to observe the physical results of a mutation. The white eye color was later found to be caused by a **mutation**, or change in the genetic code. The change in DNA sequence for the wild-type red eye color resulted in a change in the mRNA transcript and a change in the resultant protein. Thus, red pigment could not be produced, and the lack of color gives the eyes the white appearance.

Mutations come in different forms and have different results. **Somatic cell mutations** are those that occur in the DNA of a body cell, potentially during replication. Somatic cell mutations have the potential to cause harm in the individual but pose no risk to any offspring. **Germ cell mutations**, on the other hand, occur in the DNA of cells that produce gametes. Thus, any offspring created from these gametes will inherit the mutations, while the individual initially possessing the mutation will not be affected. In some cases, the expression of the germ cell mutation is so harmful that the offspring does not survive; this is described as a **lethal mutation**.

Often mutations involve a change in just a single nucleotide within a gene sequence. This **point mutation** can happen in one of three ways. When a different nucleotide replaces the original, a **substitution** has taken place. In other cases, a nucleotide is added to or removed from the original sequence, resulting in an **insertion** or **deletion**, respectively. Both insertions and deletions are characterized as **frameshift mutations**, those that affect the way the DNA triplets are arranged into mRNA codons and thus more dramatically alter the specific sequence of amino acids that are found in the final protein.

In other cases, the mutation involves a much larger segment of a chromosome, potentially affecting several genes (see Figure 9.2). If a piece of a chromosome breaks off and is lost, the mutation is described as a **deletion**. When a segment of a chromosome breaks off, inverts itself directionally, and then reinserts itself along the same region of the chromosome, the mutation is characterized as an **inversion**. Other chromosome mutations involve **translocation**, when a segment of a chromosome breaks off and relocates itself to another, non-homologous chromosome. Many mutations happen spontaneously, some as a result of replication errors and others due to exposure to **mutagens**, substances in the environment like cigarette smoke or UV light that can cause mutation.

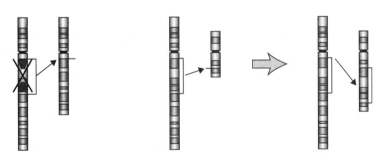

**Figure 9.2** Chromosome Mutations: Deletion and Translocation

One type of chromosome mutation is much larger in scope than those described above. A **nondisjunction** mutation occurs during meiosis when the chromosomes fail to separate and travel toward opposite poles. In the end, some gametes will contain one extra chromosome, and others will have one too few. This can result in Down syndrome when there is an extra 21st chromosome or other conditions such as Turner syndrome or Klinefelter syndrome when the sex chromosomes are affected.

# Human Genetic Traits and Disorders

One of the most common traits tested in humans is the ABO blood type. Most people are familiar with traits like A-positive or O-negative to describe blood type. Those descriptions actually include two phenotypes, one for ABO blood type (A, B, AB, or O) and one for the Rh factor (positive or negative). Here we will focus on the ABO blood-type component specifically, a trait with a fairly complex expression on its own.

The ABO blood type is considered a **multiple-allele trait** because it has more than the typical two alleles per gene. There are three alleles in this case, $I^A$, $I^B$, and $i$. The alleles for blood type A ($I^A$) and for type B ($I^B$) are codominant to each other and dominant over the allele for type O ($i$). Thus, two genotypes exist that result in the type A phenotype ($I^A I^A$ and $I^A i$), and similarly two exist that produce the type B phenotype ($I^B I^B$ and $I^B i$). There is only one genotype that creates the O phenotype ($ii$) and one that creates the AB phenotype ($I^A I^B$).

Many physical traits of interest to humans are also inherited in a complex fashion and are thus more challenging to predict. Characteristics like height, skin color, and eye color are all **polygenic**, having many separate genes involved that collectively influence expression of the phenotype. Other traits, like the genetic disorder Huntington's disease, are considered **single-allele traits**. Huntington's disease is the result of a unique protein produced from the inheritance of a mutated allele. It is unfortunately not only fatal and currently incurable, but it has a late onset and thus an individual with the condition may have already had children and passed it on without knowing it.

The inheritance pattern of yet other traits in humans is described as sex-influenced. As in the case of pattern baldness, **sex-influenced traits** are carried on autosomes (unlike sex-linked traits) but produce phenotypes influenced by the presence of sex hormones. The higher levels of testosterone present in males cause the baldness allele to be expressed as dominant, while the relatively lower levels of the hormone in females produce a recessive allele in females. It is thus less likely for females to show the baldness trait, for they have to inherit an allele for the condition from each parent. Males, on the other hand, only need to inherit one allele from either parent.

# Biotechnologies and Future Applications

It seems as if everyone in modern times is to some degree familiar with the concept of DNA fingerprinting, a means of establishing genetic identity utilized not only in forensic science but also for establishing paternity, verifying the species of a fossil, and various other applications.

To create a DNA fingerprint, a series of different protocols are carried out in sequence. First, a technique called the **polymerase chain reaction (PCR)** is performed on each sample to create potentially billions of exact copies of the specific DNA under investigation, and it does so by simulating the natural process of DNA replication. The exponential amplification of the quantity of DNA in the samples will allow for visualization of DNA later, as the multiple copies will collectively create a macroscopic band on the fingerprint.

Second, **restriction enzymes** are used in each DNA sample to cut the long DNA strands into smaller, more manageable fragments. The size of each fragment depends on the specific action of the restriction enzyme employed. Restriction enzymes are naturally occurring in bacteria as a means of digesting foreign viral DNA but have been isolated and synthesized for use in biotechnology. Here, a particular restriction enzyme recognizes a given repetitive sequence of nucleotides in DNA and cleaves the molecule at that site. Thus, individuals with unique DNA sequences will produce some fragments of different lengths and can be distinguished from one another on the final fingerprint.

Now, each sample of DNA up for comparison is placed into a tiny well within a solidified gel. The gel contains pores small enough to allow for the separation of DNA fragments of different size once the DNA begins migrating out of the well and down the gel. This movement of DNA is produced by applying an electric current to the gel. DNA is negatively charged in solution due to

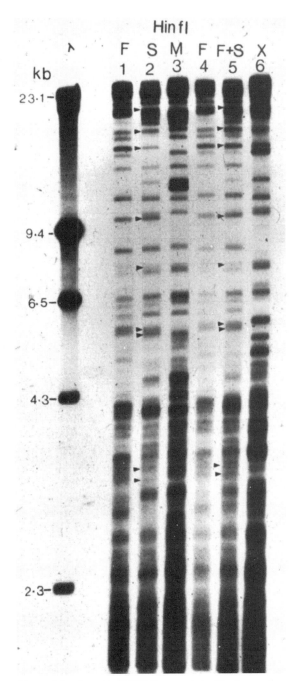

Figure 9.3a An actual DNA fingerprint with six results for banding pattern comparison. The lane on the left (lambda) is shown for a relative size comparison in kilobases (kb).

https://upioad.wikimedia.org/wikipedia/commons/8/8f/The_first_DNA_Fingerprinting_picture_reported_in_Indian_judiciary.jpg

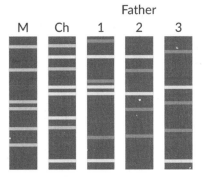

Figure 9.3b Idealized results of a DNA fingerprint comparing genes from the known mother (M) and child (Ch) with three possible fathers.

https://commons.wikimedia.org/wiki/File:DNA_paternity_testing_en.svg

Create a caption for Figure 9.3b to explain which of the three possible fathers is actually the biological father given the fingerprints of the known mother and child.

its many phosphate groups, so a negative charge is applied near the well to push on the like-charged DNA, and a positive electrode is placed on the opposite end to pull the DNA in that direction. The current is stopped before the smallest fragments of DNA reach the bottom of the gel, signifying the end of the **gel electrophoresis** process.

Finally, the DNA is stained so that all of the bands are as clearly visible as possible, and the resultant DNA fingerprint is then analyzed (see Figure 9.3). If this fingerprint were being used to try to match an unknown blood sample found at a crime scene with DNA from possible suspects, then only if a suspect's DNA sample exactly matched that from the crime scene would they remain under investigation.

Many other applications of biotechnology exist today, and it seems as if more new innovations are emerging all the time. Some of these, like **gene cloning** and **gene therapy**, are more specifically considered **bioengineering** because both techniques involve the manipulation of naturally occurring DNA sequences into new forms for expression.

In gene cloning, a restriction enzyme is used to cut open a bacterial plasmid and to cut out a healthy human gene from the genome, like the one to make insulin. The restriction enzymes used here must be a type that leaves **sticky ends** on the DNA molecule when it cleaves. These jagged overhangs are considered "sticky" because they are regions where hydrogen bonds could re-form should the appropriate complementary sequence be present. When the human gene is introduced to the open bacterial plasmid, the ends of the gene are complementary to the ends of the plasmid. Another enzyme called **ligase** is used to seal up the strands, and then that plasmid is ready for use in a bacterium. It is now considered to comprise **recombinant DNA**, DNA containing genes from the genomes of different species.

Bacteria like *E. coli* are then placed in conditions that encourage them to undergo transformation and pick up the recombinant DNA. When they do, the bacteria are now considered **transgenic** organisms. When they transcribe and translate the plasmid, they make insulin for use by people suffering from diabetes. When they replicate the plasmid during asexual reproduction, they clone the gene and pass on the instructions to all of the cells in the resultant colony.

A relatively new biotechnology known as CRISPR (clustered regularly interspaced palindromic repeats) has incredible potential to accelerate the field of biotechnology even further. These DNA sequences naturally occur in bacteria as a defense mechanism against viral infection, but researchers have harnessed that power in an effort to "correct" the mistakes represented by mutations that are responsible for a plethora of human disorders and diseases.

**EXERCISE**
**9·1**

**Vocabulary Building.** *Provide a definition for each of the following vocabulary terms. When possible, identify any roots in the term and use them to help create the definition.*

1. sex-linked trait

_____

_____

2. point mutation

_____

_____

3. polygenic trait

_____

_____

4. DNA fingerprint

_____

_____

5. restriction enzyme

_____

_____

**EXERCISE**
**9·2**

**Multiple Choice.** *Select the best response from the options provided to answer each question or to complete each statement.*

1. A mutation that can affect an individual but not its offspring is considered what type?

a. germ cell mutation         c. somatic cell mutation

b. lethal mutation             d. point mutation

2. All of the following are examples of chromosome mutations *except*

     a. deletion                  c. translocation

     b. nondisjunction        d. substitution

3. Which of the following is an example of a sex-influenced trait in humans?

     a. eye color                c. red-green color blindness

     b. hemophilia            d. pattern baldness

4. The process by which fragments of DNA from a sample can be separated based on size is called

     a. DNA fingerprinting      c. PCR

     b. gel electrophoresis      d. gene therapy

5. Which of the following is the *least* relevant to gene cloning?

     a. PCR                   c. sticky ends

     b. bacterial plasmid       d. recombinant DNA

---

**EXERCISE**
**9·3**

**Short Answer.** *Write brief responses to the following.*

1. Must a substitution mutation produce a noticeable effect in the resultant protein? Explain.

_____

_____

_____

_____

_____

_____

2. How is a sex-linked trait similar to a sex-influenced trait? In what key ways are they different?

_____

_____

_____

_____

_____

_____

3. Briefly outline the steps required to produce a DNA fingerprint.

_____

_____

_____

_____

_____

_____

**Interpreting Diagrams.** *Examine the following figure representing the transmission of the ABO blood-type gene from parents to offspring. Use the information in the figure to answer the questions that follow.*

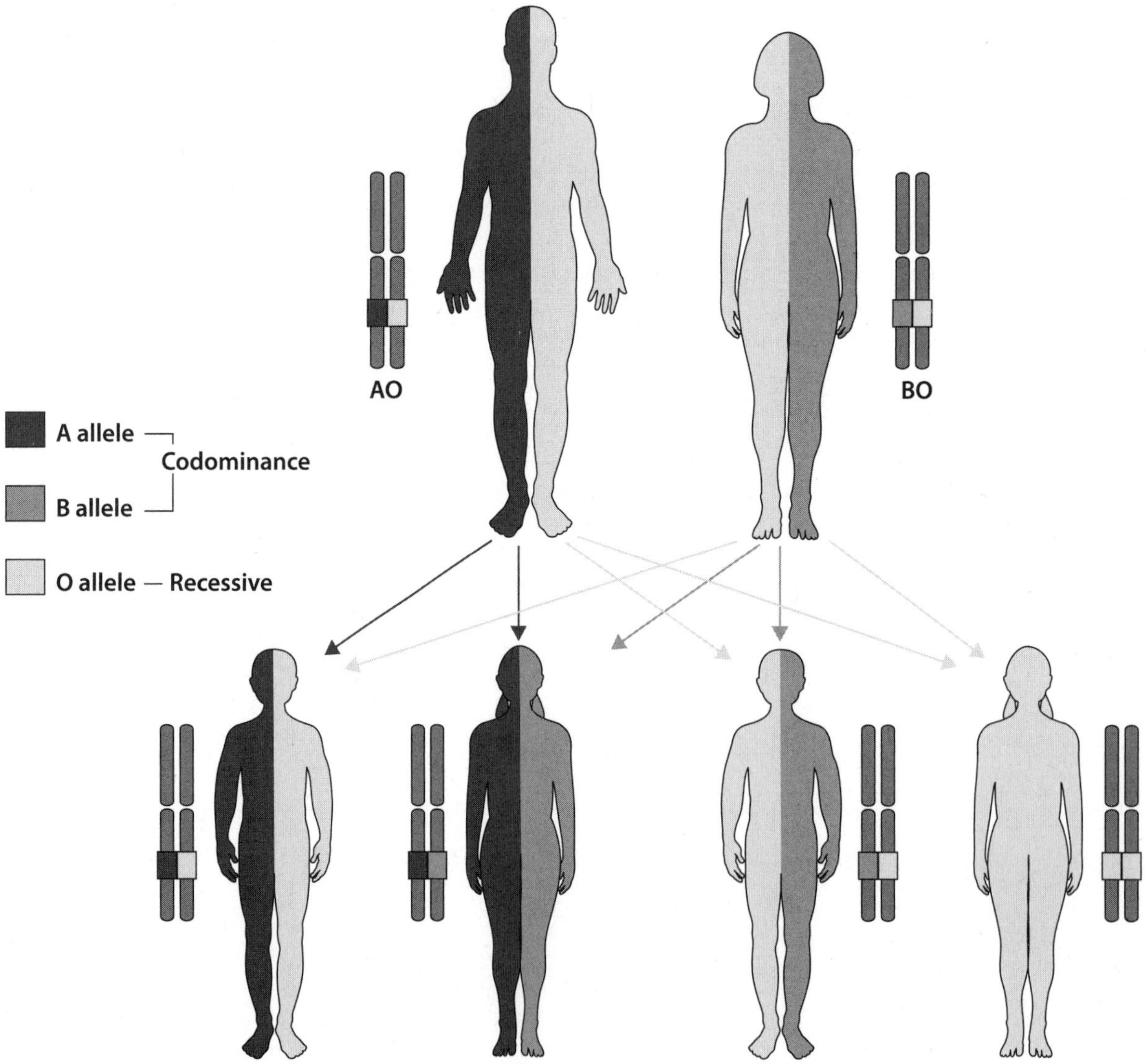

AO

BO

A allele ⎤
           Codominance
B allele ⎦

O allele — Recessive

1. What phenotype does the mother possess? The father?

_____

2. What genotypes are possible among any offspring produced from these parents?

_____

3. What is the probability of this couple producing a child with a heterozygous genotype?

_____

4. What is the probability of this couple producing a child with type O blood?

_____

**EXERCISE
9·5**

**Thinking Thematically.** *For each of the following themes of biology, choose a different concept from this chapter and explain how it provides a useful illustration of that theme.*

1. natural interdependence

_____

_____

_____

_____

_____

_____

2. science methodologies and applications to society

_____

_____

_____

_____

_____

_____

3. form facilitates function

_____

_____

_____

_____

## For Further Investigation

If you can gather enough information, you can determine your genotype for the ABO blood-type gene. If you already know your blood type, then you are a step ahead. If you are type AB or type O, then what genotype must you have? If you are type A or B, then you know you have at least one allele of that type, but what could the other allele potentially be? To determine the unknown allele or to try to determine your blood type if it is unknown to you, investigate the blood types of your parents and/or your offspring and see if that helps you reach a final conclusion.

# EVOLUTION

# The Origin and Evolution of Life on Earth

·10·

For most of modern human history, the way in which life originated on Earth was a topic of wild speculation and was generally misunderstood. This was true not only for how people thought about how life is generated and perpetuated but also for the mechanism and timeline by which the first life appeared on Earth. Before the Scientific Revolution, the vast majority of humans believed in **creationism**, the notion that a higher power generated all life on Earth as it exists in its present form. Most also adhered to the concept of **spontaneous generation**, the belief that life could spontaneously take form if an inanimate object were struck with a "vital force." Once scientists began investigating this issue further, new hypotheses began to emerge that incited controversy in the scientific community and among the general population. Quite a few related topics remain controversial to some into the present day.

## Continuity of Life

Well into the 1600s, spontaneous generation was the way in which most people explained how new nonhuman life arose. When food left out eventually would spoil, this new living growth on obviously dead food was explained away as spontaneous generation. When maggots emerged from meat left unattended, again spontaneous generation must have been at work. No evidence of another organism's presence was observable at the time, so what else could explain it?

Over the course of some two hundred years, three different scientists are credited with testing new hypotheses that eventually led to the discrediting of the notion of spontaneous generation. First, seventeenth-century Italian naturalist and physician Francesco Redi attempted to disprove spontaneous generation by testing rotting meat in jars for the presence or absence of live maggots. The key to Redi's experiment was that he covered some jars with a fine mesh such that air could pass but macroscopic particles were prevented from entering. In any uncovered jar with rotting meat, maggots eventually appeared as expected. In the mesh-covered, meat-filled jars, however, no maggots were observed, and the meat appeared unchanged throughout the same time period. While Redi believed that this was enough evidence to refute spontaneous generation altogether, many people still believed that for "lesser life" like bacteria and fungi, spontaneous generation still applied.

Almost a century later, another scientist attempted the same effort as Redi. Italian biologist and priest Lazzaro Spallanzani experimented with microorganisms and organic broth, a soupy mixture of macromolecules. He hypothesized that microorganisms, too tiny to be seen, resided in the air and were responsible for much of the "spontaneous generation" still apparently observed. Spallanzani boiled organic broth in separate flasks. Only one of the flasks was left uncorked; the other was securely stoppered to prevent any air flow. Similar to the results observed by Redi, Spallanzani observed plentiful growth of bacteria and fungi in the uncorked flask but no growth in the corked flask. Again, Spallanzani believed that his results were enough to refute spontaneous generation. But critics suggested that the boiling of the broth also killed off the vital force in the flasks, and then any new vital force was prevented from entering the stoppered flask, thus preventing spontaneous generation from taking its hold.

Another century later, an already well-known French scientist, Louis Pasteur, was determined to refute spontaneous generation once and for all. He took the basic principles of Spallanzani's experimental design but incorporated some ingenious changes (see Figure 10.1). He used only one flask with boiled broth, and this time the flask had a long, curved neck. Pasteur hypothesized that the convoluted neck of the flask would trap microorganisms present in the air before reaching the flask, but the uncorked design would certainly allow air (and any vital force, if present) to reach the broth. For a year, the broth remained unchanged, but after Pasteur then broke off the neck of the flask to create a straight neck instead, abundant growth occurred in the broth in only a few days. Finally, enough evidence had been amassed to abandon any belief in spontaneous generation and instead embrace the concept of **biogenesis**, the idea that only life can create life. But that immediately begs the question, what was responsible for the creation of the first life on Earth?

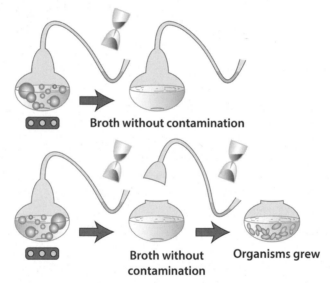

**Broth without contamination**

**Broth without contamination**

**Organisms grew**

**Figure 10.1** Pasteur's Experiment

# Earth, Prelife

According to accumulated geological evidence, the Earth began its formation some 4.6 billion years ago (bya). The early Earth was tumultuous, constantly bombarded with meteorites and astronomical debris and experiencing extremes in heat. The surface likely melted several times before cooling permanently some 4 bya, and it is assumed that after that point, organic matter could accumulate in such a way as to potentially support life.

These benchmark dates in the Earth's history have been determined primarily through the systematic determination of the age of the sediment that makes up the Earth itself. This process is more specifically known as **radiometric dating**, and it involves an understanding of radioactive isotopes, or radioisotopes, and their various, predictable patterns of decay. A **radioactive isotope** decays by emitting a neutron from its atomic nucleus, and in doing so, it loses mass. This predictable loss of mass over time allows scientists to experimentally determine each radioisotope's unique **half-life**, the length of time required for a radioactive sample to lose half of its initial mass. Another essential piece of information is the relative abundance of the radioisotope in nature relative to its more common and stable version. Armed with this data, the **absolute age** of a sample, its age in years (within an error range), can be determined.

When the Earth first formed, it certainly was not hospitable to any form of life that we currently understand. Volcanic eruptions were common, extreme heat continued, and toxic gases filled the atmosphere. It wasn't until the Earth began to experience a significant cooling that temperatures approached the range in which water can exist in its three very useful states, gaseous water vapor, liquid water, and frozen ice. Oceans and other bodies of water were formed, preparing at least some of the Earth's surface for life.

# From an Inorganic World to Cellular Life

In the 1920s, Russian biochemist Alexander Oparin hypothesized that simple organic monomers can be generated from inorganic ones if under appropriate conditions with a significant energy input. It wasn't until some thirty years later that Oparin's hypothesis was tested by others in what is now known as the **Miller-Urey experiment**. American chemists Stanley Miller and Harold Urey devised an apparatus that simulated a set of conditions likely on the early Earth, mixing hydrogen gas ($H_2$), methane ($CH_4$), ammonia ($NH_3$), and water vapor and sparking the gaseous mixture with energy to simulate a natural source like lightning. Results demonstrated that various small organic molecules like amino acids were generated in the experiment, indicating that the essential chemical building blocks of life could be generated from a previously inorganic world (see Figure 10.2).

Once simple organic molecules are present, small cell-like assemblages begin to spontaneously form due to their chemical properties, much like soap bubbles forming from detergent and water. Called **microspheres** and **coacervates** depending upon their specific composition, these structures clearly show how the first cells likely came about. Certainly countless of these assemblages formed that were not "fit" for their environment and thus degraded. When one by chance possessed a useful set of characteristics that proved successful in the environment, that **protocell** would have a greater chance of "surviving" than the other assemblages. According to current evidence, it took at least one billion years for life to take hold in true cellular form. The oldest bacterial fossils date back to at least 3.5 bya.

Simple prokaryotic life flourished for nearly 2 billion years, and new traits evolved in different lineages. Some bacteria known as **cyanobacteria** evolved the ability to photosynthesize; their oxygen by-product significantly altered the composition of the atmosphere. Ongoing oxygen production eventually led to the production of the **ozone layer**, an event that allowed life to move from water and now inhabit land.

Between 1.5 and 2 bya, events known as **endosymbiosis** were responsible for creating a new eukaryotic form of life. When one large prokaryote engulfed a smaller one, it maintained the structure and the function of that now-internal prokaryote. The large prokaryote had really become a eukaryote, having internal, membrane-bound organelles. Substantial evidence indicates that the chloroplast evolved from endosymbiosis of cyanobacteria, while the mitochondrion similarly evolved from early aerobic prokaryotes.

# History of Evolutionary Theory

Once life existed on Earth, how did it really go about changing and evolving into different species? Most people immediately and definitively associate Charles Darwin with the concept of

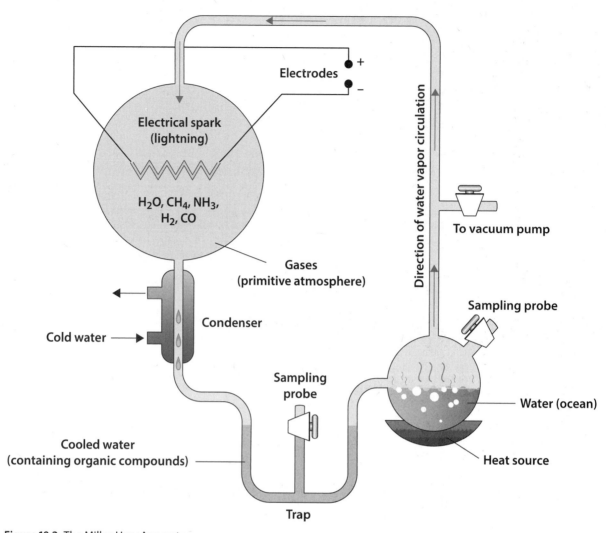

**Electrodes**
+
−

**Electrical spark
(lightning)**

$H_2O$, $CH_4$, $NH_3$,
$H_2$, CO

Direction of water vapor circulation

To vacuum pump

**Gases
(primitive atmosphere)**

**Sampling probe**

**Condenser**

**Cold water**

**Sampling
probe**

**Water (ocean)**

**Cooled water
(containing organic compounds)**

**Heat source**

**Trap**

**Figure 10.2** The Miller-Urey Apparatus

Create a caption for Figure 10.2 by explaining the significance of the products created from such simple reactants using the Miller-Urey apparatus.

evolution. Few know the whole story, however, of another young naturalist and budding evolutionary biologist named Alfred Wallace, who independently developed a theory of evolution strikingly similar to that of Darwin's and actually formally proposed it jointly with Darwin. Darwin had decidedly developed his theory earlier than Wallace, for Wallace was just a young lad in the 1830s when Darwin began collecting data on the HMS *Beagle* tour that circumnavigated the globe and brought him to the famed and inspirational Galapagos Islands. Darwin's first known attempt at visually explaining the relatedness of life was recorded on the ship in what became the first known phylogenetic tree (see Figure 10.3).

Darwin was not the first scientist to suggest evolution, for even Darwin's grandfather Erasmus also believed in the concept. What Darwin (and Wallace) is credited for is uncovering the mechanism by which evolution occurs: **natural selection**. Darwin proposed that at the heart of natural selection is the natural variation in traits that exists among members of the same species. Some traits confer an advantage to survival and/or reproduction such that possessing those traits is associated with a disproportionate ability to pass them along to the next generation. This is described as **evolutionary fitness**.

Darwin began to truly understand evolution upon analysis of the massive collection of specimens he encountered while acting as a naturalist on the *Beagle*. Most famous are his analyses of

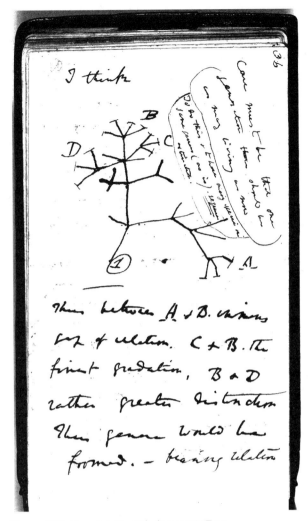

**Figure 10.3** Darwin's First Phylogenetic Tree

the variety of finches that he found on the Galapagos. As he traveled from island to island, he observed unique species on each island, seemingly related to the particular food sources available. Some finches had large, rounded bills suitable for picking up large seeds from the ground, others possessed long, pointed beaks for pulling insects out of tiny holes, while yet others had sharp bills structured for sucking blood from mammals. What Darwin was observing were the various **adaptations** that different populations evolved over time to become better suited for their environments. Darwin was also beginning to understand the idea of **adaptive radiation**, the pattern of species becoming increasingly unique as they travel further away from the ancestral species.

What Darwin did not know at the time is that evolutionary change of all types is, at the most fundamental level, caused by genetic change. As a spontaneous mutation occurs and changes the sequence of nucleotides within a gene, the resultant protein produced through transcription and translation may carry that change forward in the form of a new trait. The fitness of the trait is tested in a specific environment; traits associated with increased fitness tend to be preserved because the organisms survive to reproduce and pass on those traits, while those that are unfit are strongly selected against.

Although Darwin and Wallace jointly share credit for development of the theory of evolution, Darwin will forever be remembered as the father of evolutionary theory and believed by many to be the most significant figure historically within the field of biology.

# Evidence and Patterns of Evolution

Early evidence of evolution primarily takes the form of **fossils**, the preserved remains of long-dead organisms (see Figure 10.4). Most fossils tend to preserve best in **sedimentary rock**, which forms as layers of rock experience extreme pressure, dehydrate, and become more compact.

Individual fossils or their surrounding sediment can be tested and their absolute age determined when sufficient radioactive materials are found. Scientists also use the various sedimentary layers, or **strata**, to determine the relative age of fossil remains. According to the **law of superposition**, a lower stratum is older than a stratum positioned higher up in the rock because sediment is continually deposited and builds up over time.

More recently, increased explorations and study, along with significant advances in technology, have allowed for new evidence for evolution to be uncovered. Evolutionary biologists look for **homologous structures** as evidence of a relatively recent common ancestor (see Figure 10.5). These are traits whose genes are derived from a similar gene possessed by the ancestor species but that have been modified over time and may produce very different-seeming phenotypes. For example, the underlying forearm structure of all mammals is very similar, but the forearm may be positioned within the wing of a bat, the flipper of a dolphin, or the arm of a human. Homologous structures are often associated with **divergent evolution**, the pattern observed when closely related species become increasingly unique as they continue to evolve away from the ancestral species as in adaptive radiation (refer to Figure 10.3).

Other structures might at first appear similar between species, but upon closer inspection, the similarities are found to be mostly superficial. These traits are instead termed **analogous structures** and are evidence not of a close common ancestor as with homologous structures, but instead likely evidence that the species faced similar environmental pressures and happened to come up with similar adaptations. Examples include the wing of an insect and the wing of a bird; the wing trait was arrived at independently in each evolutionary lineage. Analogous structures are associated with **convergent evolution**, a pattern observed when species from different lineages evolve in such a way that makes them appear increasingly similar over time.

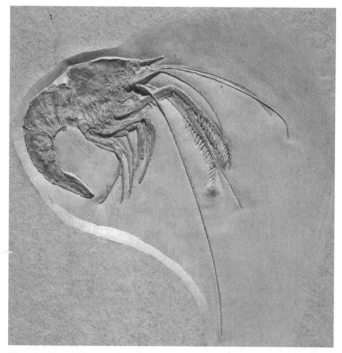

**Figure 10.4** A Shrimp Fossil

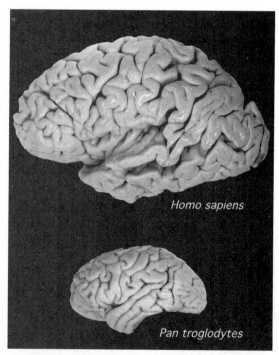

**Figure 10.5** Homologous Structures: Human and Chimp Brains

**Vestigial structures**, those that have lost their function in the present species and have usually become significantly reduced in size, can help to reinforce evolutionary relationships when compared to other species that retain a functional version of the trait. The human appendix and whale pelvis are classic examples of such vestigial structures. **Comparative embryology** is also useful in establishing evolutionary relationships. For example, the early embryos of all vertebrates are strikingly similar, as are the patterns of development that vertebrates follow when achieving a more mature form. Only when different types of appendages begin to take form (e.g., wings in a duck, legs in a salamander, and fins in fish) do the embryos become obviously distinct.

Most recently, **comparative biochemistry** is proving to be very informative of evolutionary relationships. Complete genomes of separate species can be compared for like nucleotide sequences in a technique called **DNA-DNA hybridization**. Using this method, it has been verified that humans and chimpanzees are at least 98 percent genetically identical, firmly rooting the chimp as our closest living relative. Protein sequences can also be compared; the more amino acids shared within a given protein, the more closely related the species.

# Microevolution

The most specific way of assessing whether there is evidence that evolution is occurring within a population is to take a microevolutionary approach. The very smallest level at which evolution can occur is at the allele level, for alleles determine genotype, which determines phenotype. Because **evolution** acts directly on phenotypes, it is most specifically defined as the change in an **allele frequency** in a population over time.

The first individuals to view evolution from this perspective were actually not biologists at all. An English mathematician, G. H. Hardy, and a German physician, Wilhelm Weinberg, each independently developed the same mathematical model for assessing microevolution, which later came to be known as the **Hardy-Weinberg equilibrium**.

The underlying principle of the Hardy-Weinberg equilibrium is that, in order to maintain equilibrium and for the allele frequencies to stay stable, five conditions must be held true: no net migration, mutation, or selection can be occurring; random mating should be taking place; and the population should be statistically large. If any of those assumptions are violated, then allele frequencies can change and evolution would then occur.

From another perspective, if allele frequencies are changing and microevolution is thus occurring, it is due to **gene flow** as individuals enter or leave the population; a mutation that creates a new allele for the trait; or one of the alleles having an adaptive advantage over the other. If individuals are participating in **assortative mating**, selecting mates based on like appearance or close proximity, or if they are selecting mates based on certain traits or behaviors, then allele frequencies may be changing due to **sexual selection**. Finally, if the population is very small, then chance events alone can actually result in the change of allele frequency. This phenomenon is called **genetic drift**.

To calculate allele frequencies in a particular population, information about the **gene pool**, the total genetic information available in the population, is needed. The frequency of the dominant allele for a trait is calculated by dividing the total number of dominant alleles measured in the gene pool by the total number of alleles present for that trait in the gene pool. The frequency of the recessive allele can be determined in the same fashion. This approach to assessing microevolution is called **population genetics**. If the allele frequencies for a given trait are changing over time, that is evidence that **microevolution** is occurring.

# Macroevolution

From the largest perspective, evolution can result in **speciation**, the formation of a new species. Generally speaking, this **macroevolution** results from the accumulation of enough new traits over time to make these organisms a distinct group from the ancestral species. But what really makes a species a species?

Just as when classification began and taxonomists utilized a morphological approach, biologists used to determine a new species based primarily upon a significant change in external appearance or body form. Although other species concepts exist, the one that most biologists agree upon today is called the **biological species concept**, which states that members of a species are defined by the ability of individual members to reproduce with each other and produce fertile offspring. Although useful for most organisms, this concept is admittedly not equally applicable to all subjects of study in biology, including organisms that reproduce asexually and those that are extinct.

While countless different scenarios exist that might lead to speciation, the central mechanism tends to involve some form of isolation from the original population. Much of the time it is **geographic isolation**, the physical separation of a segment of the population over time. If the environments that groups from the same species are in are different enough in ways that impact survival and/or reproduction, then the groups may undergo speciation. To verify speciation, organisms from each group are interbred. If fertile organisms can no longer be produced, then speciation has in fact occurred.

**Reproductive isolation** often helps to maintain the integrity of separate species over time by preventing successful reproduction. Many **prezygotic isolating mechanisms** exist that prevent a fertilized egg from ever forming correctly. One example is **temporal isolation**, observed when the timing of breeding seasons differs enough to prevent interbreeding between different species.

Other postzygotic isolating mechanisms may also be at work. While not able to prevent a **hybrid** zygote (a fertilized egg produced from sperm and egg from different species) from forming, these mechanisms usually involve dysfunctional development of the embryo to such an extent that, if an offspring organism can be produced, it has a vastly shortened life span. In the most extreme sense, true hybrid organisms can be produced that seem to develop normally and live a relatively normal life, like the mule produced from a horse-donkey mating, but these hybrid organisms are sterile and unable to contribute their genes from there in an evolutionary sense.

Speciation has been observed to take place at different paces in different contexts or with different types of organisms. Two basic patterns exist, gradualism and punctuated equilibrium. **Gradualism**, although still relevant today, was (it is no longer assumed to be the only pattern) the assumed rate of speciation from the beginnings of evolutionary theory and Darwin's notion of descent with modification. Gradualism states that new species may form very slowly and steadily over long periods of time. Later evidence suggested that another pace of speciation also exists, **punctuated equilibrium**. In this case, evolution maintains species for very long periods of time but then rapidly establishes new species that can differ quite markedly. Both gradualism and punctuated equilibrium rely upon the notion that an ancestral species can accumulate new traits over time to eventually become a new species; it is mainly the relative rate of the accumulation that differs. Over the entirety of evolutionary history, however, the significance of this difference is arguable and varies widely depending on the species under consideration. In the end, any speciation is really dictated by the pace of environmental change.

**Vocabulary Building.** *Explain the relationship between the following pairs of vocabulary terms.*

1. spontaneous generation, biogenesis

_____

_____

2. relative age, absolute age

_____

_____

3. homologous structure, divergent evolution

_____

_____

4. evolution, natural selection

_____

_____

5. gradualism, punctuated equilibrium

_____

_____

EXERCISE
10·2

**Multiple Choice.** *Select the best response from the options provided to answer each question or to complete each statement.*

1. Natural selection acts directly on
   - a. phenotype
   - b. alleles
   - c. genotype
   - d. both a and b

2. The smallest level of biological organization in which evolution can occur is the
   - a. individual organism
   - b. population
   - c. cell
   - d. species

3. The wing of a butterfly and the wing of a bat are considered what type of structures?
   - a. homologous
   - b. vestigial
   - c. analogous
   - d. divergent

4. Which of the following would *not* be expected to contribute to a change in allele frequency over time?

   a. immigration

   b. large population size

   c. artificial selection

   d. mutation

5. By which of the following mechanisms can the creation of a new species occur?

   a. postzygotic isolation

   b. reproductive isolation

   c. geographic isolation

   d. all of the above

EXERCISE

**10·3**

**Short Answer.** *Write brief responses to the following.*

1. Explain how the law of superposition and relative dating are useful in understanding the age of fossils.

   _____

   _____

   _____

   _____

   _____

   _____

2. Some bacteria have recently been discovered to possess the ability to digest plastic and nylon, both synthetic substances created by humans. Outline the steps that would have to take place for these strains of bacteria to evolve from an ancestral species without that ability.

   _____

   _____

   _____

   _____

   _____

   _____

3. Compare and contrast microevolution and macroevolution.

   _____

   _____

   _____

   _____

EXERCISE

# 10·4

**Interpreting Diagrams.** *Use the information in the following graphs demonstrating rates of speciation to answer the questions that follow.*

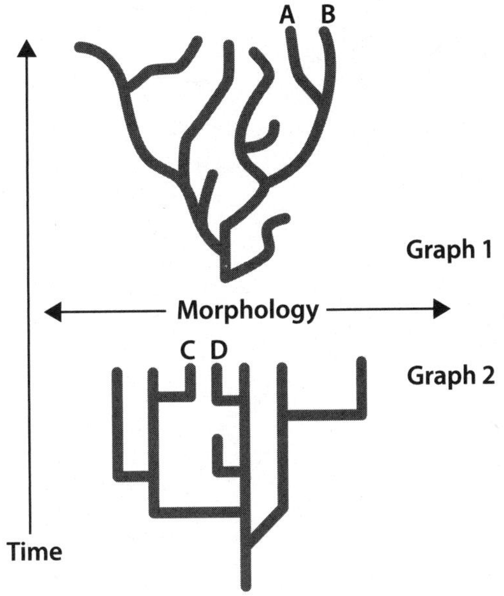
A B
Graph 1
Morphology
C D
Graph 2
Time

1. Which graph demonstrates punctuated equilibrium? ______________________
2. Which graph demonstrates gradualism? ______________________
3. Which species demonstrate divergence? ______________________
4. Which species demonstrate convergence? ______________________

EXERCISE

# 10·5

**Thinking Thematically.** *Choose a concept from this chapter and explain how it provides a useful illustration of the following themes of biology.*

1. continuity and change

2. science methodologies and applications to society

_____

_____

_____

_____

_____

3. regulation and feedback

_____

_____

_____

_____

_____

## For Further Investigation

Visit a natural history museum, either in person or online, and carefully observe different fossil specimens. First, employ your knowledge of fossil types (e.g., cast, petrification) to classify different fossils. Then, find an example of homologous structures and an example of analogous structures among the fossils observed. Do some quick Internet research as needed to verify the degree of relatedness amongst the specimens under comparison.

# Unifying the Diversity of Life

## Classification

As you might imagine, attempting to systematically name the millions of different species on the planet is no easy task. Although about 2 million species have formally been classified to date, biologists estimate that between 10 million and 50 million species exist, and the number could even be much higher. Early efforts at classification relied primarily on an organism's outward appearance, as evidenced by the now outdated morphological species concept discussed in Chapter 10. Some of the earliest approaches to taxonomy on record were made in the third century B.C.E. by Aristotle, Greek philosopher and student of medicine. As was common practice at the time, Aristotle first separated humans from the rest of life in a hierarchical sense and then subdivided the remainder into either plants or animals. Some of his further subdivisions of the plant and animal groups are still useful today, but his classification scheme certainly is not sufficient for describing the vast and diverse microbial life unseen and thus unknown at the time, and fails to recognize important macroscopic groups of fungi and protists as distinct from their plant and animal counterparts. **Taxonomy**, the scientific approach to naming and classifying all biological organisms, is significant also in that it clearly demonstrates that science is a process equally dependent on the current technology, prior knowledge, and social values of the time.

## Linnaeus, the Father of Modern Taxonomy

After Aristotle, the next significant advances in taxonomy weren't made until the 1700s when Carolus Linnaeus, a Swedish zoologist and botanist, devised the system of **binomial nomenclature** still in use today. Linnaeus believed that two names, the genus and the species, were needed to correctly name and identify an organism. To ensure consistency in communication of the scientific name of an organism, the genus name is capitalized, the species name is not, and the entire name is italicized in print or underlined if handwritten. The scientific name for humans is thus *Homo sapiens*, sometimes also written with the genus abbreviated as *H. sapiens*. Often our **subspecies** name is also included to differentiate modern humans from our early human ancestors; in that case, we are *H. sapiens sapiens*, while they would be *H. sapiens archaic*. Subspecies are also used to indicate geographically isolated populations of like species or artificially selected breeds.

The genus and species were just two **taxa**, or levels, of a more complex, multilevel classification system (see Figure 11.1). Linnaeus devised the hierarchical system such that the **kingdom** was the most inclusive (or least specific) taxon, which was further divided into phyla. Each **phylum** was separated into classes, each **class** into orders, each **order** into families, each **family** into genera, and finally

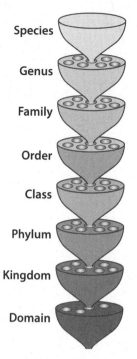

**Figure 11.1** Classification Levels

each **genus** into species. The **species** taxon represented the least inclusive and most specific level of classification and was intended by Linnaeus to uniquely describe all organisms on the planet. Although the numbers and names of many taxa continue to change throughout the years as new evidence is gathered and discoveries are made, the framework of Linnaeus's taxonomic scheme still guides the naming of organisms today.

Until recently, Linnaeus's seven-level system was widely utilized for a classification scheme that recognized five kingdoms of life. While this may have fit the needs of taxonomists for some time, the extreme biodiversity that has come to light since the time of Linnaeus has necessitated the addition of supplementary taxa like subspecies or even infraorder. The traditional five kingdoms have become six, and molecular research has provided evidence to suggest that it is useful to add a taxon more inclusive than kingdom, the **domain**.

To date, there are three domains of life that have been described. Domain Archaea includes all of kingdom Archaebacteria, the most ancient life forms on Earth. Similarly, kingdom Eubacteria is the exclusive member of domain Bacteria. Both Archaea and Bacteria are composed of unicellular prokaryotes that reproduce asexually, but the similarities really end there. This is not unreasonable considering that they have had more than 3 billion years of time for evolutionary divergence. The third domain, Eukarya, includes the four eukaryotic kingdoms of life: Protista, Plantae, Fungi, and Animalia. Most protists are unicellular, although there are notable multicellular species. Protists also show extreme diversity in metabolism and habitat. Plants are multicellular, photosynthetic autotrophs, while animals are multicellular heterotrophs. Fungi are mostly multicellular but have several significant unicellular varieties. They are exclusively heterotrophic, like animals.

## Modern Classification

A newer classification approach, called **systematics**, attempts to update the Linnaean criteria for organization into different taxa. Instead of relying solely on morphological and ecological similarities, systematicists also utilize molecular evidence and homologies in embryonic development to establish natural taxonomic relationships. Close analysis of the early embryonic development of different vertebrate species indicates overwhelming similarities, even when the end organisms might be as different as a salamander and a giraffe. DNA-DNA hybridization, as mentioned in Chapter 10, is a technique used to estimate the number of shared base pairs between different species; humans and chimpanzees have been shown to be at least 98 percent genetically identical. Examination of amino acid sequences in important proteins like hemoglobin from both species have only reinforced the DNA hybridization results, indicating only very few differences in a hemoglobin molecule containing 574 amino acids. Most systematicists agree that modern classification schemes should represent **phylogeny**, the evolutionary history of organisms. As shown in Figure 11.2, these relationships are shown graphically through a **phylogenetic tree** (or phylogenetic diagram).

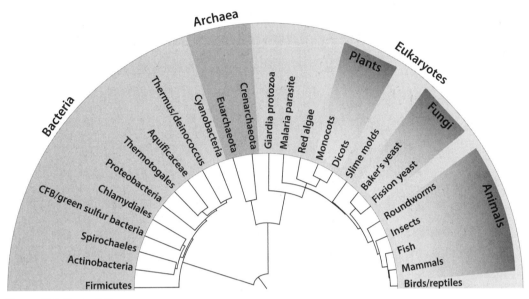

**Figure 11.2** Tree of Life

Create a caption for Figure 11.2 to explain where the members of kingdom Protista are specifically situated within the phylogeny and what this placement suggests about their relatedness.

**Cladistics** is a specific method of phylogenetic analysis that involves assessing the number of shared, derived characters among different organisms. **Derived characters** are those that evolved only within the smaller group of organisms being considered, thus any shared, derived characters are inferred to be inherited from a common ancestor, and organisms possessing them make up a **clade**. A **cladogram** is the graphical representation of the relatedness of the clades under analysis.

EXERCISE
**11·1**

**Vocabulary Building.** *Provide a definition for each of the following vocabulary terms. When possible, identify any roots in the term and use them to help create the definition.*

1. binomial nomenclature

_____

_____

2. taxonomy

_____

_____

3. phylogenetic tree

_____

_____

_____

**Multiple Choice.** *Select the best response from the options provided to answer each question or to complete each statement.*

1. Aristotle's very early classification system would not accurately describe which of the following types of organisms?
   a. fungi
   b. protists
   c. bacteria
   d. all of the above

2. Which of the following taxa is inclusive of all the others?
   a. genus
   b. domain
   c. order
   d. kingdom

3. The most ancestral of the six modern kingdoms of life is
   a. Eubacteria
   b. Monera
   c. Protista
   d. Archaebacteria

4. Of the six modern kingdoms of life, the one most closely related to animals is
   a. Fungi
   b. Protista
   c. Eubacteria
   d. Plantae

5. An organism that is unicellular and eukaryotic could possibly be a(n)
   a. animal
   b. protist
   c. bacterium
   d. plant

**Short Answer.** *Write brief responses to the following.*

1. Why does the field of taxonomy continue to change over time? Does this mean that the current methods employed for classification are unreliable? Explain.

_____

_____

_____

_____

_____

_____

2. What is the key feature used to separate domain Eukarya from the other two domains? What characteristics are used to distinguish domain Archaea from domain Bacteria?

_____

_____

_____

_____

_____

_____

3. Modern systematics considers many factors when classifying a newly observed organism. List several of such factors that are currently employed to classify that would not have been used by either Aristotle or Linnaeus.

_____

_____

_____

_____

_____

_____

_____

**EXERCISE**
**11·4**

**Interpreting Diagrams.** *Examine the following phylogenetic tree. Use the information in the figure to answer the questions that follow.*

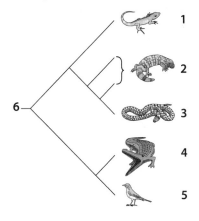

1. Which species is depicted as the ancestor of all others?    _____

2. Of species 1, 2, and 3, which two appear to be the most closely related?    _____

3. Which living species is most closely related to 4?    _____

4. Which species, 1 or 5, should share more traits with species 3?    _____

**EXERCISE**
**11·5**

**Thinking Thematically.** *For each of the following themes of biology, choose a different concept from this chapter and explain how it provides a useful illustration of that theme.*

1. form facilitates function

_____

_____

_____

_____

_____

_____

2. continuity and change

_____

_____

_____

_____

_____

_____

3. science methodologies and applications to society

_____

_____

_____

_____

_____

_____

_____

## For Further Investigation

For a favorite organism of your choosing, use the Internet to determine the names of the appropriate taxa in the eight-level scheme. Do you recognize any of the names? Can you think of other organisms that belong to any of the same taxa? Internet sites like the Encyclopedia of Life (eol.org) are particularly useful.

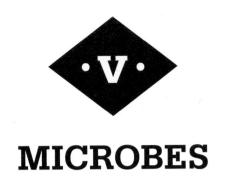

# MICROBES

# Straddling the Living World

## Viruses

We encounter **viruses** every day, some relatively benign like those that cause the common cold or influenza, and others that are widely feared, such as those responsible for HIV, Ebola, and most recently, COVID-19 viruses, too tiny to be seen under a light microscope, mostly evaded intense scientific study until the invention of the electron microscope in the 1930s. Their activity had been observed before that time; microbiologists had observed living cells literally dying before their eyes due to an assumed viral infection. Just before the turn of the twentieth century, the tobacco mosaic virus became the first officially recognized pathogen, although it took decades to actually be recognized as a virus. What exactly constituted a virus remained to be fully understood for quite some time, and to this day, some debate continues regarding their exact classification.

## Not Quite Life

Why is it that viruses are tough to classify? We are familiar with the way that many viruses can cause sickness not only in our own species but in many others, so our first inclination may be to assume they are living things, just like bacteria or trees or worms. When examined more closely, however, only hints of evidence of the characteristics of life are present. Most convincingly, viruses are **acellular**, composed of not even a single cell. In fact, many viruses contain only DNA or RNA for genetic instructions wrapped up in a protective protein coat. There is no cellular membrane, no ribosomes, and no other typical organelles to speak of. For that reason alone, a virus is distinct from all true forms of life, but the justifications only continue from there. Viruses have no ability to metabolize, detect and respond to stimuli, and grow or develop. The one characteristic of life that viruses do seem to possess, reproduction, is a task that can only be carried out within the confines of its **host** cell. In fact, viruses are **obligate intracellular parasites**, particles that must depend on resources within a host cell for replication.

## Viral Replication in Prokaryotes

Because the main concern of every virus is finding a way to infect a host and make more virus, its structure is well designed to suit this purpose. If the host cell is a prokaryotic bacterium like *E. coli*, then the virus is referred to as a **bacteriophage**. Given that the host cell lacks a nucleus and keeps its own DNA out in the cytoplasm, the virus simply has to have a means of inserting its nucleic acid through the outer coverings of the bacterium (the cell membrane and cell wall, when present), and the viral DNA is then ready to be accessed in the bacterial cytoplasm. For this reason, bacteriophages are fairly simple in structure. Their outer protein coat, called a **capsid**, protects the nucleic acid contained inside and acts as a vessel to transport the genetic instructions from one host and insert them in the next.

Once the bacteriophage has docked down on the surface of its prokaryotic host, it uses its collar and tail structures to insert its nucleic acid through the bacterial cell wall and membrane and into the cytoplasm. There, it either is immediately transcribed and translated by the bacterium—unknowingly using its own enzymes and resources to help further the infection—or it inserts itself into the circular bacterial DNA and takes cover for some time. If the first option is employed, the **lytic cycle**, the continual production of new virus, ultimately leads to the eventual **lysis**, or bursting, of the cell (see Figure 12.1). Although the original host cell dies in the process, it releases many new copies of virus to now infect similar neighboring cells. The exponential production of new virus through replication and lysis continues until the source of new host cells runs dry.

If the infected prokaryotic cell does not immediately begin production of new virus, then the **lysogenic cycle** is instead underway. During this latent phase, the viral DNA becomes inserted into the main bacterial chromosome amidst the bacterial genes; at this point, the viral DNA is now called **prophage DNA**. When the bacterial cell undergoes reproduction, it copies the entirety of its chromosome, including the prophage DNA, and passes along these genetic instructions to all offspring cells that result. Eventually, an environmental trigger will encourage the prophage DNA to excise itself from the bacterial chromosome, at which point it then enters the lytic cycle and begins the active expression of new virus.

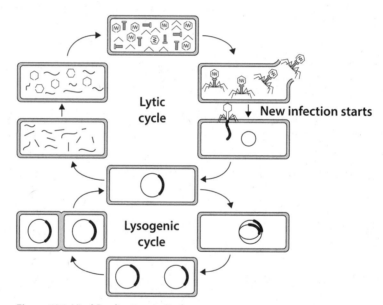

**Figure 12.1** Viral Replication in Prokaryotes

Create a caption for Figure 12.1 to summarize the lytic and lysogenic cycles of viral replication.

# Viral Replication in Eukaryotes

Of much more direct concern to humans certainly is the way in which viruses infect eukaryotes like protists, fungi, plants, and animals like us. Instead of simply injecting their genetic instructions as would a bacteriophage, these viruses have a more complex interaction with their eukaryotic host cell. The virus itself is typically more complex, often possessing an **envelope** exterior to the capsid. This envelope is composed of the lipid bilayer from a previous host cell's nuclear or cell membrane, and it aids the virus in host cell recognition and infection.

Viruses of eukaryotes usually trick the host cell into voluntarily engulfing the viral particle, and in doing so, the viral capsid is delivered into the host cell's cytoplasm. From there, the DNA housed within the capsid is directed into the nucleus, where it becomes incorporated into the host cell's chromosome and called a **provirus**. It will eventually be transcribed and translated along with the bacterial DNA, but in this case, the result is new viral particles. Examples of DNA viruses include the chickenpox virus and HPV, the virus that causes genital warts and increases the chances of the human host developing cervical cancer.

Some viruses do not contain DNA at all and are collectively termed RNA viruses. Once in the cytoplasm, the RNA can be used as a template to make viral proteins directly. A subset of the RNA viruses, the **retroviruses**, has a unique mechanism of infection. They come prepackaged not only with genetic instructions but also with copies of an enzyme called **reverse transcriptase**. Once incorporated into the cytoplasm of the host cell and released from the capsid, the RNA is acted upon by the enzyme and used as a template to make viral DNA. It thus reverses the direction of the information flow within the central dogma. Once the viral DNA is made, it then acts like a DNA virus; the DNA becomes a provirus and eventually leads to viral protein synthesis and eventual assembly into new viral particles. The virus that causes rabies is a simple RNA virus; retroviruses specifically include influenza, the virus that causes the flu, and HIV, the virus that leads to AIDS (see Figure 12.2).

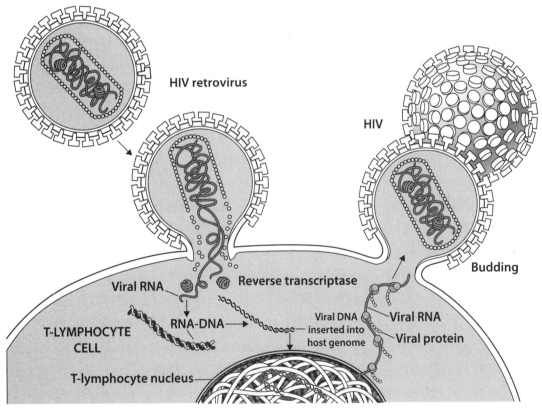

**Figure 12.2** HIV Infection

Treating viruses can be tricky; it is impossible to "kill" something that isn't truly living in the first place. Many can remain dormant for a very long time, both inside and outside of the host cell, depending on the virus and the environmental circumstances. RNA viruses remain a particular challenge, for their single-stranded genetic instructions are less stable and more subject to mutation. When mutation changes the structure of the outer receptor molecules of the virus, any vaccine designed to prevent it will lose its effectiveness. A **vaccine** acts like a primer for the immune system, for it exposes immune cells to important pieces of the virus involved in recognition and infection of host cells. If a vaccinated individual is later exposed to that pathogen, then his or her immune system is effectively prepared with antibodies to fight off the invader.

Some of the vaccines engineered to combat COVID-19 were unique in that they contained mRNA instructions necessary for creating antibodies specific to the coronavirus. In this way, they used a shortcut to the central dogma, providing a transcript ready for translation into antibody proteins. No actual viral genes were required.

EXERCISE
12·1

**Vocabulary Building.** *Provide a definition for each of the following vocabulary terms. When possible, identify any roots in the term and use them to help create the definition.*

1. acellular

_____

_____

2. bacteriophage

_____

_____

3. provirus

_____

_____

4. vaccine

_____

_____

EXERCISE
12·2

**Multiple Choice.** *Select the best response from the options provided to answer each question or to complete each statement.*

1. Viruses are not classified into any kingdom of life because they lack
   a. DNA
   b. proteins
   c. cellular organization
   d. a capsid

2. A bacteriophage infects an *E. coli* bacterium, and the cell soon bursts, releasing many new viral particles. This suggests that this particular virus is
   a. latent
   b. in the lytic cycle
   c. lysogenic
   d. a prophage

3. The lipid bilayer that surrounds many viruses is called a(n)

    a. envelope                            c. provirus

    b. capsid                            d. membrane

4. If a virus is classified as a retrovirus, then it must contain

    a. RNA                              c. reverse transcriptase

    b. DNA                              d. both a and c

5. A provirus can be found in the

    a. host cell's nucleus                c. cytoplasm of the bacterial host

    b. viral capsid                      d. all of the above

**EXERCISE 12·3**

**Short Answer.** *Write brief responses to the following.*

1. Compare and contrast the lytic and lysogenic cycles.

_____

_____

_____

_____

_____

_____

2. How do viruses cause infection in eukaryotic cells? Name a DNA virus and an RNA virus that affect human populations. Explain how these two viruses affect human populations differently. How is the specific nucleic acid contained significant?

_____

_____

_____

_____

_____

3. How are vaccines effective in preventing viral infections? What can be done if an infection has already occurred? Use your knowledge of the global response to control COVID-19 to situate your response.

_____

_____

_____

_____
_____
_____

EXERCISE
**12·4**

**Labeling Diagrams.** *Fill in the blanks using the following terms to correctly label the diagram representing a virus that infects eukaryotic cells.*

capsid

envelope

nucleic acid

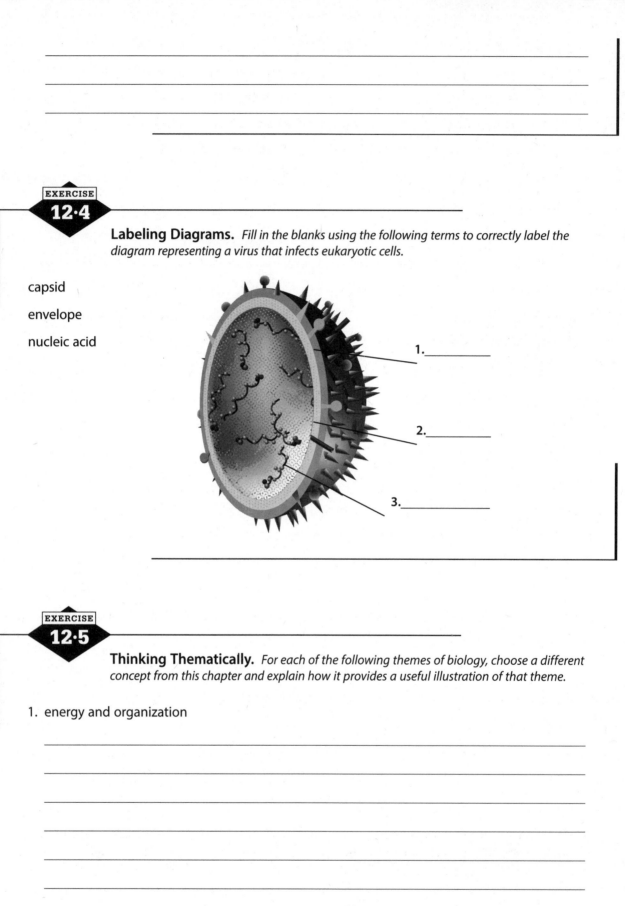

1._____

2._____

3._____

EXERCISE
**12·5**

**Thinking Thematically.** *For each of the following themes of biology, choose a different concept from this chapter and explain how it provides a useful illustration of that theme.*

1. energy and organization

_____
_____
_____
_____
_____
_____

2. form facilitates function

_____

_____

_____

_____

_____

3. continuity and change

_____

_____

_____

_____

_____

_____

## For Further Investigation

Some viruses are being used in biotechnologies in a process called *gene therapy*, using the infectious nature of the particle to "infect" cells with healthy genes. Use the Internet to research the technique and some current applications of the therapy.

# Prokaryotes Old and New

## Kingdoms Archaebacteria and Eubacteria

Like viruses, prokaryotes are much too small to be seen with the naked eye and occupy virtually every known habitat on Earth. In fact, there are more bacteria living on and in each human than that human has of their own cells. Most of them tend to get a bad rap, guilty by association with the relatively few pathogenic bacteria out there. What most people fail to recognize, and often have a tough time coming to terms with, is that the vast majority of bacteria are either helpful or neutral in their interactions with humans and other organisms. They carry out numerous significant ecological roles and perform essential ecosystem services. Some are used by humans to help process foods and create biomedical therapies. Others have particular evolutionary significance, and many are currently under investigation in attempts to uncover applications for sustainable energy and **bioremediation**, natural waste management and environmental clean-up. While all prokaryotes used to be grouped into a common kingdom, Monera, more recent phylogenetic evidence has led to the replacement of Monera by the newer kingdoms **Archaebacteria** and **Eubacteria** (discussed in Chapter 11).

## Ancient Microbes

Recall that the first forms of life on Earth most closely resembled organisms in domain Archaea today. Archaea tend to be **extremophiles**, organisms that thrive in environments considered extreme by most other life forms. Today, what we find extreme was actually typical to the early Earth; now these environments are restricted to relatively few isolated places. Three primary groups of Archaebacteria are currently recognized, each distinguishable from the others by the favored type of extreme environment. **Halophiles** prefer very salty conditions like the Dead Sea (so named before extremophiles were found to exist there) and the Great Salt Lake. **Thermophiles** instead favor extremes in heat, like bacteria that inhabit thermal hot springs, volcanoes, and deep-sea vents. Finally, the **methanogens** live in anaerobic conditions like the guts of termites, cows, and humans, using available substances like hydrogen gas ($H_2$) and carbon dioxide ($CO_2$) for energy and releasing methane gas ($CH_4$) as a metabolic by-product.

Archaebacteria are all unicellular, asexual prokaryotes. They possess an outer cell wall, although the exact composition varies among species. The cell wall lacks the peptidoglycan characteristic of other bacteria and often includes rare lipids. The cell membrane of Archaebacteria is also structurally distinct from Eubacteria, for the exact composition of lipids and of embedded proteins varies between the kingdoms. Genetically, Archaeans possess **introns** (interrupting sequences that must be spliced out before translation) in their genes, linking them more closely to eukaryotes and distinguishing them from the other prokaryotes.

The sequence of their rRNA and the resultant ribosomal structure is also more similar to that of all eukaryotes than to the Eubacteria. The cellular structure of Archaebacteria is otherwise relatively simple, usually possessing a single, circular chromosome, ribosomes, and cytosol. Recall that all prokaryotes lack a nucleus and other membrane-bound organelles; this holds true for Archaeans.

# Modern Bacteria

Most prokaryotes that humans are familiar with belong to domain Bacteria and kingdom Eubacteria. These bacteria are always unicellular, asexual prokaryotes, much like Archaea. They are distinct, however, in that Eubacteria always have cell walls composed of **peptidoglycan**, a molecule that is both protein and carbohydrate in nature. The modern bacteria also lack introns in their genes, so transcription of the DNA and translation of the mRNA can be carried out simultaneously. Their DNA is circular in form, as if one end of the double helix attached to the other, and constitutes the **main chromosome**. This circular conformation helps prevent degradation of the molecule in the absence of a protective nucleus.

Bacteria also possess numerous smaller circular chains of DNA known as **plasmids**. These carry nonessential genes that may confer a selective advantage to a bacterium, e.g., antibiotic resistance, and can sometimes be exchanged between different bacteria. They are also very useful in bioengineering, including the cloning of the insulin gene to help treat people with diabetes.

Many bacteria possess an outer **capsule** external to their cell wall that helps to prevent dehydration, and if pathogenic, also from being acted upon by the host cell's immune action. When this capsule is made up of fuzzy sugars that help the bacterium attach to other cells, it is specifically called a **glycocalyx**. A **pilus** is another structure that can help a bacterium attach to a surface, but pili are long, hairlike extensions made of proteins.

A specialized type of pilus, called a *sex pilus*, is used to transfer genetic information found on a bacterial plasmid to another bacterium in a process known as **conjugation**. Conjugation thus allows the bacteria to increase genetic diversity in their population in spite of their restriction to **binary fission**, a form of asexual reproduction in which a bacterium copies its DNA and then divides its cytoplasm to create two genetically identical offspring cells.

Other means of increasing genetic diversity during the nonreproductive life of bacteria include **transformation**, the process of a bacterium picking up naked DNA from its immediate surroundings, and **transduction**, the process of a bacteriophage transferring genetic information between bacteria as an unintended consequence of its infectious lifestyle.

The cells of a eubacterium are typically spherical (**coccus**), rod-shaped (**bacillus**), or spiral (**spirillum**) in form (see Figure 13.1). If attached and organized into long filaments, the spherical bacteria are termed **streptococci**; the same type of bacteria arranged into clusters are collectively called **staphylococci**.

Eubacteria are classified into smaller subgroups often by means of a laboratory technique known as the Gram stain. When this particular dye is applied to certain bacterial cells, their cell walls appear a deep purple-blue under the microscope; these are known as **Gram-positive** and possess a thick layer of peptidoglycan in their cell walls. Other eubacterial cells have instead a thin layer of peptidoglycan covered by another membrane; these stain a lighter pink and are known as **Gram-negative**. The Gram stain classification not only makes it easy to differentiate bacteria in a sample under the microscope, but it also helps direct medical treatment for an infection. Gram-positive and Gram-negative bacteria are susceptible to different

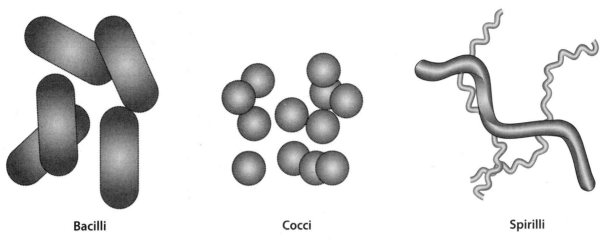

| Bacilli | Cocci | Spirilli |

**Figure 13.1** Bacterial Forms

antibiotics and utilize very different cellular mechanisms to produce the familiar symptoms of that infection.

Eubacteria demonstrate diversity in metabolic styles that are directly or indirectly related to the divergent habitats they occupy. **Obligate anaerobes** (like the methanogens from kingdom Archaebacteria) find an oxygen-containing environment toxic, whereas **facultative anaerobes** are more flexible and can live in the presence or absence of oxygen. **Obligate aerobes**, on the other hand, must be in an oxygenated environment. Eubacteria that are obligate anaerobes are likely to be **chemoautotrophs**, meaning that they use inorganic carbon dioxide ($CO_2$) to make their own organic food, but instead of using energy from the sun to accomplish this task, they utilize energy from inorganic molecules in the environment. Some bacteria are **photoheterotrophs**, relying on the sun for an energy input and on consumption of organic molecules for a carbon source. Most common eubacteria are either **photoautotrophs**, organisms that use the sun's energy to build food from inorganic $CO_2$, or **chemoheterotrophs**, organisms that need to eat to obtain energy and organic carbon. (For comparison, plants are photoautotrophs and animals are chemoheterotrophs.)

# Bacteria and the Environment

Modern thermophiles that occupy deep-sea vents constitute the basis for benthic food chains, as their chemosynthetic metabolism uses geothermal energy and simple inorganic molecules to make organic food in the ocean depths where the sun's radiant energy certainly cannot reach. Others are recyclers along with the fungi and some invertebrate animals, helping keep our ecosystem clear of dead organic matter. Yet other bacteria are **nitrogen-fixers**, converting otherwise unusable nitrogen gas ($N_2$) into forms that can be incorporated by all other organisms within the ecosystem. Many of these bacteria inhabit the soil, and some live symbiotically within plant roots.

Many people are surprised to learn of the numerous ways that we have found to utilize bacteria to our benefit. Dairy products are created with the help of lactose-fermenting bacteria, live bacterial cultures are an ingredient in yogurt, and other prokaryotes are involved in the production of many cheeses, pickled items, and soy products. Many enzymes have been isolated from bacteria that are useful in detergents and in biotechnologies, in the processing of petroleum, and the cleanup of oil spills.

Humans certainly find many forms of bacteria a nuisance or maybe even a major threat. Pathogenic species often are not intentionally causing the symptoms that characterize the infection but rather are releasing metabolic wastes into their environment in an attempt to maintain homeostasis. The problem really occurs when that environment is our bloodstream, intestinal tract, or brain. Some Gram-positive bacteria release **exotoxins** directly into their environment while metabolizing. Examples include *Streptococcus*, the culprit that causes strep throat, and *Bacillus anthracis*, the cause of anthrax. Other Gram-negative bacteria don't release their symptom-inducing **endotoxin** until the cell dies, thus potentially causing symptoms long after the bacteria themselves have been killed. Examples include *E. coli* and the related *Salmonella*, strains of both of which cause serious intestinal disease and are often contracted through contaminated food.

While the immune system may be able to quell a bacterial infection on its own, often we employ **antibiotics** to encourage the process. Many different antibiotics exist today that utilize varied mechanisms to destroy bacterial cells; some are naturally derived from other kingdoms of life, while others are synthetic. While the intentions behind antibiotic use are generally good, the vast overuse and misuse medically, agriculturally, and industrially have led to **antibiotic resistance** in many strains of bacteria. Just as insect populations quickly develop resistance to insecticides, antibiotics can suddenly lose their effectiveness when their target bacteria show similar signs of resistance.

Resistance occurs through the exchange of an **R plasmid** during conjugation and natural selection of bacteria in a population with favorable traits. Those bacteria that happen to possess the R plasmid have genes that protect them from the given antibiotic, so they will survive while all others will die. Thus, the only bacteria that survive to reproduce and establish the next generation are those that possess the R plasmid and that are resistant. This can be a big problem when some dangerous bacteria develop resistance to multiple strains of antibiotics, as observed in the potentially fatal MRSA (methicillin-resistant *Staphylococcus aureus*). Proper use of antibiotics, including avoiding their use when a viral infection is actually present and using the drugs only and exactly in accordance with the prescription, can help slow this growing problem.

## EXERCISE 13·1

**Vocabulary Building.** *Define each of the following vocabulary terms and provide an example for each.*

1. extremophile

_____

_____

2. Gram-positive bacterium

_____

_____

3. antibiotic

_____

_____

## EXERCISE 13·2

**Multiple Choice.** *Select the best response from the options provided to answer each question or to complete each statement.*

1. A microbiology student examines cells from a bacterial culture under the microscope. She observes spherical bacteria arranged in a chain. Her preliminary classification should be

    a. staphylococcus

    b. spirochetes

    c. streptococcus

    d. bacillus

2. Which of the following typical bacterial structures is most related to symptoms felt during an infection by a pathogenic bacterium?

    a. pilus

    b. exotoxin

    c. cell wall

    d. plasmid

3. A bacterium that makes its own food from inorganic molecules but does *not* rely on the sun's energy to do so is called a

    a. photoheterotroph

    b. chemoautotroph

    c. photoautotroph

    d. chemoheterotroph

4. If a prokaryote were observed to contain peptidoglycan in its cell wall, it would definitely be classified as a(n)

    a. eubacterium

    b. Gram-positive bacterium

    c. archaebacterium

    d. both a and b

5. If a bacterium becomes resistant to an antibiotic, which of the following structures would be involved?

    a. plasmid

    b. endospore

    c. flagellum

    d. capsule

## EXERCISE 13·3

**Short Answer.** *Write brief responses to the following.*

1. What evidence is there to suggest that members of Archaebacteria are more closely related to all members of domain Eukarya than to the other prokaryotic kingdom, Eubacteria?

_____

_____

_____

_____

_____

_____

2. Describe the role that humans play in contributing to antibiotic resistance.

_____

_____

_____

_____

_____

3. Argue that bacteria are more helpful than harmful to humans overall.

_____

_____

_____

_____

_____

_____

**Labeling Diagrams.** *Fill in the blanks using the following terms to correctly label the diagram representing a typical eubacterium.*

cell wall

pilus

plasmid

ribosome

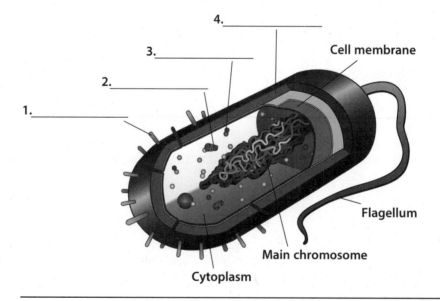

4. _____

3. _____

2. _____

1. _____

Cell membrane

Flagellum

Main chromosome

Cytoplasm

**Thinking Thematically.** *For each of the following themes of biology, choose a different concept from this chapter and explain how it provides a useful illustration of that theme.*

1. continuity and change

_____

_____

_____

_____

_____

_____

2. form facilitates function

_____

_____

_____

_____

_____

_____

3. science methodologies and applications to society

_____

_____

_____

_____

_____

_____

_____

## For Further Investigation

In the United States over the past few years, there have been many reports of bacterial infection within the food supply resulting in recalls of the affected food or food product, and in many cases, sickness or even death. Conduct online research of some of these recent instances. For each case, identify the type of bacterium causing the outbreak and the means of dealing with the infection.

# ESSENTIAL AND UNDERAPPRECIATED ORGANISMS

# Simple and Diverse Life

## Kingdom Protista

Members of arguably the most diverse kingdom of life, **protists** are stuck somewhere in between the simple bacteria discussed in Chapter 13 and the more complex life in the plant, fungal, and animal kingdoms. The one unifying characteristic of protists is that they are all eukaryotic. Beyond that, protists demonstrate diversity in metabolic style, reproductive mechanisms, overall body form, and much more. Members of kingdom Protista are defined primarily by exclusion from other kingdoms, the major reason they are also the next kingdom that may be reclassified. Whether Protista will suffer the same fate as the formerly recognized kingdom Monera, only time, technology, and research will tell.

## Classification of Protists

Until ongoing research clearly establishes phylogenetic relationships among protists, it remains most practical to divide them into three broad subgroups (consider these like "super-phyla") based on their significant ecological roles (see Figure 14.1).

## Animal-Like Protists

Animal-like protists are unicellular, heterotrophic, and usually motile. In fact, the phyla of these protists are named for their means of locomotion. Probably the most well-known subgroup, **protozoa** comprises organisms that move by means of **pseudopodia**. These "false feet" are really just extensions of the cell membrane created as the protozoan forces much of its cytosol to flood one particular area of the cell in a process called **cytoplasmic streaming**. An amoeba is probably the most easily recognizable protozoa, with its irregular borders ready for the formation of pseudopodia.

    **Ciliates**, on the other hand, coordinate the beating of their cilia to direct movement in aqueous environments. The extremely common freshwater ciliate *Paramecium* has a relatively complex internal structure, an efficient feeding mechanism, and a rudimentary form of sexual reproduction. **Sarcomastigophorans** are characterized by at least one long, tail-like flagellum that helps propel them through their aqueous environment. The **apicomplexans** are the oddball of the subgroups, for they possess no form of locomotion. They have evolved to rely on animal parasitism for nutrition, so they lost the need for motility.

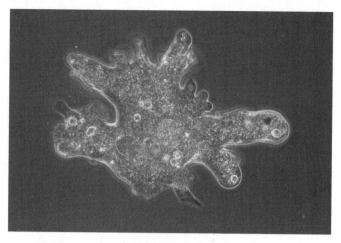
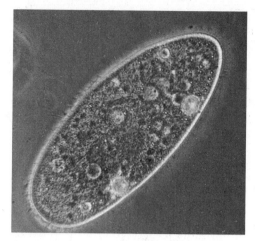
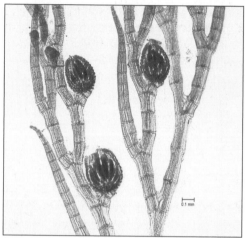
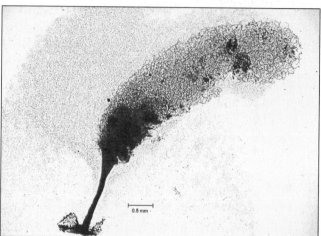

**Figure 14.1** A Variety of Protists

## Plant-Like Protists

This second group is represented mostly by the **algae**. Not only do algae contain chloroplasts rich with chlorophyll *a* and make their own food through the process of photosynthesis, but many types also store excess sugar in the polysaccharide starch and have cell walls composed of cellulose—just like plants. Algae come in a variety of colors and morphologies. Most **red algae** (phylum Rhodophyta) are marine seaweeds that can receive light at greater depths than most other algae because they possess accessory pigments called phycobilins. **Brown algae** (phylum Phaeophyta) include some of the largest seaweeds like the giant kelp; the accessory pigment fucoxanthin provides the basis for their distinct color. **Golden algae** (phylum Chrysophyta) demonstrate their characteristic color because of the high concentration of carotenoids. They are predominantly unicellular organisms that inhabit freshwater ecosystems. **Green algae** (phylum Chlorophyta) are significant evolutionarily in that the universal ancestor of all true plants belonged to this phylum. Green algae demonstrate extreme diversity, ranging in form from unicellular to colonial and multicellular varieties.

Other plant-like protists are not true algae, although they still tend to be photosynthetic autotrophs. **Euglenoids** (phylum Euglenophyta) are like green algae and plants in their collection of photosynthetic pigments but differ in that they do not have a cell wall and are exclusively unicellular. **Diatoms** (phylum Bacillariophyta) are unicellular phytoplankton that share the same accessory pigment with brown algae. They contain a unique form of sugar storage and form shell-like cell walls composed of silicon dioxide. Similarly, the **dinoflagellates** (phylum Dinoflagellata) have plated cell walls, but theirs are instead composed of cellulose. Dinoflagellates are often **bioluminescent**, possessing the ability to generate light through a chemical reaction. Many people might be familiar with the often-problematic **red tides** caused by population explosions (or **algal blooms**) of these protists. Their excessive concentration produces toxins that can negatively affect life up the food chain, including humans who may become sickened after eating infected shellfish.

## Fungus-Like Protists

This final subgroup of protists comprises two main types, the slime molds and the water molds. The **slime molds** are heterotrophic and often decomposers. Most have interesting life cycles characterized by a discrete unicellular phase and a collective, colonial-like phase that often grows to macroscopic size and helps for feeding and/or reproduction. The **water molds** are mainly unicellular parasites, often infecting fish and other aquatic organisms. One group of water molds, the **chytrids**, is hypothesized to include the ancestor of true fungi.

Although we have summarized protists here using an ecological approach, we know that is not typically the best practice in biological classification. It is done here for practical purposes, as the kingdom is so vast and diverse and the molecular research until recently has been relatively lacking. Over time, we will have a much better sense of the true phylogeny of protists.

# Protists in the Environment

Ecologically speaking, protists carry out some essential roles. The autotrophic protists produce a large amount of the atmosphere's oxygen, necessary for cell respiration in all aerobic organisms. They also constitute the foundation of most aquatic food webs, whether marine or freshwater. As just discussed, the fungus-like protists often act as important decomposers, recycling organic debris. Some members of kingdom Protista form distinct interactions with other organisms, like the symbiotic zooxanthellae, algae living inside of corals (which are animals) in reef ecosystems. Another species of protist lives inside the guts of termites, permitting them to digest the high cellulose concentrations in the wood that they consume.

Not all protists have positive interactions with other organisms; many in fact cause disease, and one in particular is responsible for approximately 1 million deaths and 250 million sicknesses each year worldwide. That culprit is the malaria-causing *Plasmodium*, which requires not only a human but also a mosquito vector to complete its complex life cycle. Also using an insect vector is the protist that causes the fatal African sleeping sickness, but this time the part-time host is the tsetse fly. Another animal-like protist, *Cryptosporidium*, causes intestinal disease in humans often through the consumption of contaminated drinking water. *Giardia* causes similar symptoms but usually is ingested from freshwater streams and lakes where it is naturally occurring.

It might be surprising to learn that protists are quite commonly used by humans in various aspects of our daily lives. As are most forms of life, some protists have proven to be a useful and nutritious food source. If you have consumed ice cream, sushi, or salad dressing, chances are you have consumed protists. The seaweeds common to sushi rolls are probably the most obvious example consumed; other foods use by-products isolated from protists, like the carbohydrate carrageenan, used in ice creams and puddings for consistency. Protist by-products are also commonly used in cosmetics, medicines, commercial detergents, and filters.

**Vocabulary Building.** *Explain the relationship between the following pairs of vocabulary terms.*

1. pseudopodia, cytoplasmic streaming

_____

_____

2. algal bloom, red tide

_____

_____

3. slime mold, water mold

_____

_____

**Multiple Choice.** *Select the best response from the options provided to answer each question or to complete each statement.*

1. All members of kingdom Protista are always
   a. unicellular
   b. prokaryotic
   c. eukaryotic
   d. asexual

2. Animal-like protists that exhibit motility include all of the following *except*
   a. ciliates
   b. protozoa
   c. sarcomastigophora
   d. apicomplexans

3. Typical algae should possess
   a. pseudopodia
   b. seeds
   c. cilia
   d. photosynthetic pigments

4. Slime molds and water molds are fungus-like in that they include
   a. recyclers
   b. parasites
   c. decomposers
   d. all of the above

5. All of the following represent helpful roles or applications of protists *except*
   a. *Plasmodium*
   b. zooxanthellae
   c. carrageenan
   d. phytoplankton

**Short Answer.** *Write brief responses to the following.*

1. Describe three ways in which members of kingdom Protista exhibit diversity.

   _____

   _____

   _____

   _____

   _____

   _____

2. *The classification of protists is often more practical than phylogenetic.* Explain the meaning and significance of this statement.

   _____

   _____

   _____

   _____

   _____

   _____

3. Consider your own diet. Which protists do you consume? Where else do you encounter protists in your everyday life?

   _____

   _____

   _____

   _____

   _____

   _____

**EXERCISE**
**14·4**

**Interpreting Diagrams.** *Examine this diagram of an unidentified protist. Use the information in the image to answer the questions that follow.*

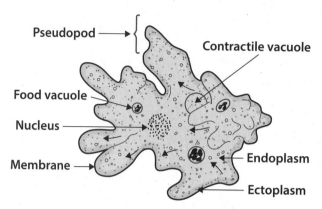

1. Which type of protist is this, animal-, plant-, or fungus-like?

2. What subgroup can it be further classified into?

3. Is this organism autotrophic or heterotrophic?

**EXERCISE**
**14·5**

**Thinking Thematically.** *For each of the following themes of biology, choose a different concept from this chapter and explain how it provides a useful illustration of that theme.*

1. science methodologies and applications to society

2. form facilitates function

_____

_____

_____

3. continuity and change

_____

_____

_____

_____

_____

_____

## For Further Investigation

Protists are everywhere but often completely overlooked. If you have access to any natural body of water, look at its boundaries and just under its surface for abundant protist life in algal form. How many different types can you observe when you look closely? Then examine some of the built world around you for evidence of protists. Examine the ingredients of your personal care products and identify some used in their production. Do some quick online research to understand what part(s) of the protist is/are used and what function it initially served to the protist.

# The Recyclers

## Kingdom Fungi

When asked to identify the largest organism on Earth, well-informed suggestions would likely include the blue whale or giant redwood. Many scientists now agree that this title belongs decidedly to a **fungus**, the honey mushroom of eastern Oregon. The size of this organism is quite staggering, occupying nearly ten square kilometers of soil, although the vast majority of the fungus resides underground and only occasionally breaches the surface with its fruiting bodies in an attempt to reproduce. These fruiting bodies are what most people would call "mushrooms" if ordering a fungus-covered pizza or salad, but the mushroom, as evidenced by the Oregon giant, is actually a much more complex organism than our diet suggests. Fungi, along with many bacteria and some invertebrate animals, constitute the majority of **saprobes** on the planet, converting dead and decaying organic matter into building blocks of vital nutrients for other organisms to use. Imagine what the world would look like without these essential and underappreciated soil dwellers—organic waste piling up everywhere and a shortage of molecules to support new and growing life.

## Characteristics of Fungi

Structurally, **fungi** can vary from unicellular to colonial to true multicellular varieties (see Figure 15.1). Regardless of their overall morphology, the cells of all fungi contain the carbohydrate **chitin** in their cell walls, store excess sugar in the form of the polysaccharide **glycogen**, and are typically haploid (containing only one pair of chromosomes per cell).

Unicellular fungal examples include **molds**, as often observed growing on our food when kept past its freshness date, and the common baking ingredient

**Figure 15.1** Sac Fungi

**yeast**. Multicellular fungi are composed of individual filamentous structures called **hyphae**. Each hypha is produced from mitosis to generate the elongated hyphae, but fungi differ regarding the internal organization. If cytokinesis follows mitosis, then the hyphae created are called **septate** because they are clearly divided by walls into individual cells (see Figure 15.2). If instead cytokinesis does not occur, then **coenocytic** hyphae are created, multinucleate structures with more continuous cytoplasm from cell to cell. Regardless of the type of hyphae present, when intertwined and functioning together within an individual fungus, hyphae form a collective structure called a **mycelium**.

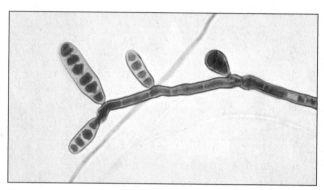

**Figure 15.2** Hypha Structure

Members of kingdom Fungi are heterotrophic but are distinguished from other heterotrophs like animals by their pattern of heterotrophy. Fungi are **absorptive heterotrophs**; their hyphae first secrete digestive enzymes into soil that contains **detritus**, dead organic matter. The enzymatic digestion occurs outside of the fungal body in the soil. Once complete, the fungus is then able to more easily absorb the nutrients across the cells' walls and into its cells. In other words, we (as animals) ingest, then digest; fungi digest, then absorb.

When it comes to reproduction, most fungi are capable of both asexual and sexual forms. Asexual reproduction often takes the form of spores produced through mitosis within specialized hyphae. Other times it takes the form of either fragmentation or budding. During **fragmentation**, a small piece of the septate fungus breaks off and grows into a new hypha and eventually an adult fungus through mitosis. **Budding** occurs in yeast when the parent cell pinches off a smaller version of itself after copying its DNA. As is always true with asexual reproduction, the offspring's cell or cells are genetically identical to those of its parent.

Sexual reproduction in fungi occurs only between opposite **mating types** called *plus* and *minus*, much like sperm is needed to fertilize the egg in plants and animals. If a hypha from a plus-type fungus fuses with a hypha from a minus-type, a genetically unique diploid structure is created that survives long enough to produce and release diverse haploid spores through meiosis. These spores are scattered away from the parent plant, each one capable of the formation of a new fungus.

# Fungus Classification

Historically, fungi were originally misclassified as plants due largely to their shared habitat and, more superficially, their similar growth patterns. Upon further investigation, it became clear that fungi are nonphotosynthetic (consider the color of mushrooms you are familiar with, and you will not think of the vibrant pigments typical for plants), and the field of **mycology** was established.

Four phyla of fungi are generally recognized, distinguished primarily by the specialized reproductive structures possessed. Phylum Ascomycota includes the **sac fungi** like yeasts, cup fungi, and *Penicillium* (the natural source of the familiar antibiotic penicillin). They reproduce sexually by means of an **ascogonium**, a structure that uses meiosis to create haploid **ascospores**. These unicellular spores will eventually be released from the parent for eventual germination into new offspring fungi.

Commonly called the **club fungi**, members of phylum Basidiomycota are named for their club-shaped reproductive structures called **basidia** found on the mushroom or mushroom-like fruiting body (see Figure 15.3). This structure is a result of the sexual reproduction between opposite mating types and eventually releases genetically unique, haploid **basidiospores** that can develop into the next generation of fungus. In a similar fashion, members of phylum Zygomycota utilize a structure called a **zygosporangium** to make their special type of **zygospores**. The final subgroup of fungi, phylum Deuteromycota, is referred to as the "imperfect fungi" because these organisms have not been observed to reproduce sexually. This subgroup is not phylogenic in nature but is rather a practical grouping for those fungi that don't fit the traditional classification scheme.

**Figure 15.3** Club Fungi

# Fungi in the Environment

As should now be clear, the primary ecological service of fungi is the nutrient recycling of dead organic matter. Some also have interesting symbiotic relationships with other organisms. The roots of many types of plants house fungal filaments interwoven between or within some of their own cells; this symbiotic association is called **mycorrhiza**. This interaction provides the plant root with increased surface area for water absorption and the fungus with sugars for food. Other types of fungi interact with either algae or cyanobacteria in a composite organism called a **lichen**. The fungus provides a safe habitat for the autotroph living just under its surface and in return receives sugar for energy.

Humans have found plenty of other uses for fungi, including but not limited to biomedical products like antibiotics and vaccines; foods like mushrooms, morels, and truffles; and food products like bread, cheese, soy products, beer, and wine.

In spite of all of the benefits the ecosystem and humans specifically receive from fungi, there are some species that interact negatively with other species. Many varieties of pathogenic fungi exist in overly moist environments; in humans, these include fungi that cause athlete's foot, vaginal yeast infections, and histoplasmosis, to name just a few. Plants also suffer from fungal infections, most of which are classified as rusts and powdery mildews. Overwatering of houseplants can lead to fungal infections of the root by providing an ideal habitat for the fungus. Inhalation of mold spores can cause allergic reactions in humans and can exacerbate the effects of asthma. Toxic and hallucinogenic mushrooms also exist; if consumed, they may cause mild to serious problems for the animal.

## EXERCISE 15·1

**Vocabulary Building.** *Provide a definition for each of the following vocabulary terms. When possible, identify any roots in the term and use them to help create the definition.*

1. saprobe

_____

_____

2. mycelium

_____

_____

3. mycorrhizae

_____

_____

## EXERCISE 15·2

**Multiple Choice.** *Select the best response from the options provided to answer each question or to complete each statement.*

1. All members of kingdom Fungi are always
   a. asexual
   b. heterotrophic
   c. multicellular
   d. prokaryotic

2. All of the following are related to asexual reproduction in fungi *except*
   a. spores
   b. budding
   c. fragmentation
   d. lichen

3. Typical fungi should possess
   a. cell walls of cellulose
   b. glucose-storing glycogen
   c. zygospores
   d. diploid cells

4. An example of a fungal pathogen with a plant host is
   a. a mold
   b. histoplasmosis
   c. *Penicillium*
   d. a rust

5. An imperfect fungus would be expected to lack
   a. spores
   b. reproductive structures
   c. mating types
   d. mycelia

**EXERCISE 15·3**

**Short Answer.** *Write brief responses to the following.*

1. Differentiate between the heterotrophy observed in fungi and in animals.

_____

_____

_____

_____

_____

_____

2. Why were fungi once misclassified as plants? What evidence suggests that they are more closely related to animals?

_____

_____

_____

_____

_____

3. Describe one positive and one negative interaction that humans have with fungi.

_____

_____

_____

_____

_____

_____

**EXERCISE 15·4**

**Interpreting Diagrams.** *Examine the following diagram representing a typical fungus. Use the information in the diagram to circle the correct answer in the statements that follow.*

1. The hyphae of this fungus are (coenocytic/septate).

2. The shape of the fungal fruiting body suggests that it is a (sac fungus/club fungus).

3. This organism is commonly called a (mushroom/yeast).

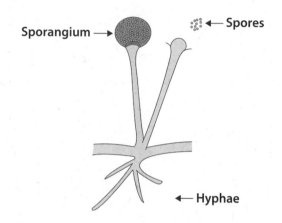

**EXERCISE 15·5**

**Thinking Thematically.** *For each of the following themes of biology, choose a different concept from this chapter and explain how it provides a useful illustration of that theme.*

1. energy and organization

_____
_____
_____
_____
_____
_____

2. natural interdependence

_____
_____
_____
_____
_____
_____
_____

3. science methodologies and applications to society

_____

_____

_____

_____

_____

_____

## For Further Investigation

Research the accidental discovery of the first antibiotic, penicillin, extracted from a fungus. What are some other significant biomedical products isolated or derived from fungi?

# The Producers

## Kingdom Plantae

About the same time that fungi began to inhabit terrestrial Earth, true land plants also evolved. In order to survive in this new environment largely devoid of liquid water, plants adapted with the evolution of important new structures. **Cuticles**, waxy outer coverings, allowed plants to retain the water they so critically need for photosynthesis even when exposed to dry air and hot sun. **Vascular tissue**, inner tubular vessels that run the length of the plant body, evolved in most plants to allow large multicellular organisms to efficiently transport water and food to and from its extremities. **Xylem** is adapted to move water and dissolved solutes from the plant roots, where it is absorbed from the soil, to the leaves, where it is needed for photosynthesis. **Phloem** is instead well structured for moving sugars from the leaves where they are synthesized throughout the plant body for energy and finally to the roots, where excess is stored.

Many plants also evolved **pollen**, sperm that lost their flagella and developed a tough, outer covering, and **seeds**, dormant plant embryos packaged with food to support growth of the eventual seedling before photosynthesis can begin. The pollen made it much easier to reproduce on land, and the seed aided in the dispersal of the offspring. Only one plant group evolved the ability to attract animal pollinators through **flowers**, but this reproductive adaptation was so successful that this group now dominates the plant kingdom. The major groups of plants classified today are shown in Figure 16.1.

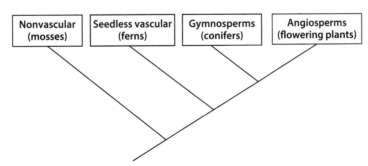

**Figure 16.1** A Simplified Plant Cladogram

Plants are essential organisms, responsible for a significant proportion of the oxygen production and carbon dioxide absorption on Earth due to their photosynthetic properties. In making their own sugar for food, they also provide the foundation of terrestrial food chains. Plants provide humans with countless ecosystem services, including food and food products from agriculture, paper materials and wood for construction, natural fabrics and dyes, medicines, and cosmetics. Look around you and on you, and you're likely to observe several of these products.

# Plant Characteristics

A defining characteristic of all plants is their life cycle pattern called **alternation of generations** (see Figure 16.2). Between the time a new plant first germinates and the time it dies, it will experience life both in a haploid phase and in a diploid phase. The haploid phase of life is called the **gametophyte** generation, while the diploid phase is termed the **sporophyte** generation.

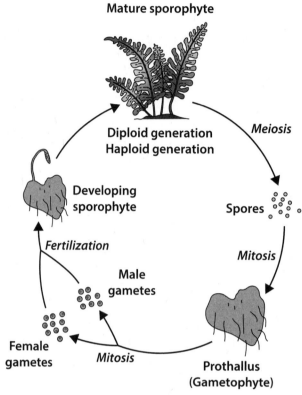

**Figure 16.2** Alternation of Generations in Ferns

# Nonvascular Plants

Nonvascular plants, those lacking any internal tubes for circulation of materials, are often collectively called the **bryophytes.** Mosses, liverworts, and hornworts make up the three major subgroups, with mosses being the most generally recognizable and characteristic of the group as a whole. Bryophytes are unique from the rest of the plant kingdom in many ways beyond the lack of vascular tissue. Most lack the true roots, stems, and leaves typical of plants and instead possess variations on the more complex structures. Mosses are very reliant on water for many of their life functions, more so than any other plant group. They are typically found growing in moist environments like shaded forest floors and near ponds and streams. Their lack of vascular tissue to move water and nutrients around internally severely limits their overall body size, as they must rely upon simple cellular transport processes like osmosis.

All nonvascular plants are also gametophyte-dominant, such that a typical moss plant contains only haploid cells with one set of chromosomes. Only when flagellated sperm from one plant are washed over to the eggs of another does fertilization occur, and the plant then creates a temporary diploid structure that represents its sporophyte generation. Using meiosis, the sporophyte creates haploid spores that it disperses into the surrounding environment. Each spore, if in the right conditions, undergoes mitosis to grow into a new, haploid, multicellular moss plant. The alternation of generations thus begins again (see Figure 16.3).

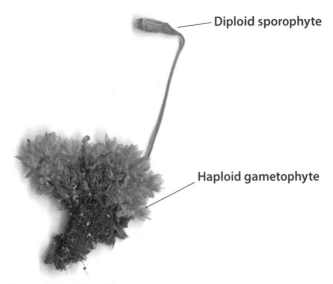

Diploid sporophyte

Haploid gametophyte

**Figure 16.3** A Moss Plant

Create a caption for Figure 16.3 by indicating the major adaptation represented by each evolutionary branch.

## Seedless Vascular Plants

The evolution of vascular tissue paved the way for the evolution of seedless vascular plants like ferns. Usually growing much larger than mosses, the seedless vascular plants actually played a predominant role in the forests during the days of the dinosaurs. Horsetails, whisk ferns, and club mosses are all subgroups, but the most recognizable are the ferns. Ferns have true roots, stems, and leaves, although some of the other subgroups lack one or more of these familiar plant organs.

All seedless plants are sporophyte-dominant, so the typical fern is composed of diploid cells with two sets of chromosomes, one received from each parent. As the name of this group implies, ferns lack seeds that are characteristic of the later groups of plants to evolve. On the underside of fern leaves, clusters of spore-producing structures called **sori** help the plant create the haploid reproductive spores for dispersal. As described with moss spores, optimal conditions will signal the fern spore to germinate in the soil via mitosis, producing an independent, multicellular gametophyte structure. If sperm from a different fern gametophyte are able to reach this gametophyte's eggs, then fertilization will occur, and a new diploid sporophyte fern will emerge from a coiled fiddlehead (see Figure 16.4).

**Figure 16.4** Fern Structures

## Forest Giants

Seeds were the next significant structural adaptation for plants, and the earlier group of seed plants to evolve possessed naked seeds. These **gymnosperms** were so reproductively successful that they quickly became the dominant autotroph in many terrestrial forests. Now, gymnosperms include the tallest and largest species on land, the giant sequoia and the coastal redwood, respectively (see Figure 16.5).

**Figure 16.5** Conifers

These uncovered seeds are often produced on the surface of hard, scaled cones, as is characteristic of the conifer subgroup. (Other less familiar subgroups of gymnosperms include the cycads, ginkgoes, and gnetophytes.) Recall that a seed is a dormant, protected plant embryo, and in that way, it represents the start of the gymnosperm's sporophyte generation. Gymnosperms are so sporophyte-dominant that the remnants of the gametophyte generation are restricted to just a few cells inside one organ of the adult plant. A mature conifer may possess both male and female cones. Within each, the reduced gametophyte is produced. In male cones, it contains the pollen, while the female gametophyte contains the ovum (egg). Wind is typically responsible for moving pollen to a female cone on another plant where pollination and eventual fertilization will occur. The female cone thus disperses the mature seed from its scales.

## Flowering Beauties

Although seeds were a tremendous evolutionary step for plants, the flowers and fruits found only within the **angiosperms** propelled this group to become the most successful plant group on the planet. This is evidenced by their biodiversity—there are at least 20,000 more angiosperm species than species from all other plant subgroups combined. Angiosperms inhabit almost every terrestrial environment on Earth, and like gymnosperms, they are exceedingly sporophyte-dominant. Two significant subgroups of flowering plants exist today, **monocots** and **dicots.** The primary distinction between these groups arises from the developmental structure of the embryo. Some angiosperms produce an embryo with one seed leaf, or cotyledon, while others produce an embryo with two cotyledons (see Figure 16.6). Monocots have flower parts in multiples of three, while dicots have them in multiples of four or five. The leaves are also distinguishable; monocots have parallel leaf veins compared to the branched, netlike veins in dicot leaves.

The secret of the flower (see Figure 16.7) is the reproductive advantage it provides its plant through the attraction of animal symbionts that assist in the pollination process. Flowers are characteristically colorful and fragrant in order to lead the pollinator directly to the male gamete, the sperm-containing pollen grain. While searching for food in the form of nectar or pollen, animals like bees and butterflies unknowingly transfer pollen from another plant.

**Figure 16.6** Angiosperms: Monocot and Dicot Flowers and Sprouts

Fruits, while not always present, provide an additional reproductive boost by encouraging animal consumption and eventual dispersal of the seeds far away from the parent plant. Seeds can only accomplish this feat by possessing a tough outer coat that resists the digestive juices in the animal's gut. Within the seed, the immature plant embryo exists along with nutritious endosperm for energy. After it is eliminated by the animal, the seed becomes deposited in the soil, where it awaits appropriate environmental conditions. Often with the influx of water and an increase in ambient temperature, the seed kick-starts mitosis, germinates and breaches the soil surface, and finally begins to rely on the sun as a metabolic energy source.

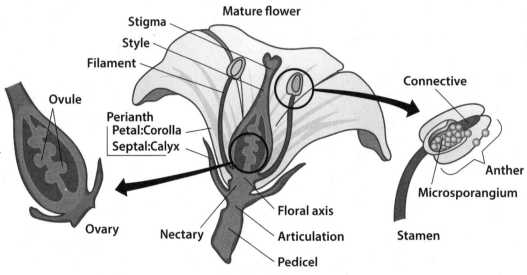

**Figure 16.7** Parts of a Flower

# Basic Plant Physiology

Most plants comprise three basic organs: roots, stems, and leaves. Roots and any derivative structures below ground make up the **root system**, one of two body systems recognized in plants. The **stems** and **leaves** that usually grow above ground together make up the other system, the **shoot system**. **Roots** are branched structures with increased surface area for maximum absorption of water and mineral nutrients from the soil. **Stems** are usually elongated structures used to prop the photosynthetic **leaves** up closer to the sun. Passive transport processes allow the plant to move these important substances into the xylem of the root without investing ATP energy. From there, adhesion and cohesion of the water molecules contributes to capillarity within the xylem tube as the water makes its way up the stem against gravity. What contributes to the majority of the water movement is actually a pull on the column of water inside the xylem from the leaf. As the stomata on the underside of a plant leaf lose water to the atmosphere, the remaining water in the tube is pulled upward. This process is referred to as **transpiration**.

The tips of stems and roots are the main sites of growth in any plant. Both contain regions of active mitosis known as **meristems**. As roots produce new cells, they push the roots down and out within the soil; when shoots produce new cells, the ends of the stem become longer and potentially closer to the sun. This type of growth is considered **primary growth**, helping the plant to more efficiently obtain essential energy and nutrients. Gymnosperms and angiosperms also demonstrate **lateral growth** that results in an increase of the girth of the stem, allowing the plant to grow much taller than it otherwise would. The specialized meristem instrumental in lateral growth is known as **cambium**. This lateral growth is familiar to most as the tree rings observable in a trunk cross section from a fallen tree or major branch.

While we usually think of plants as being immobile, they actually can adjust themselves in their environment in many different ways. Often they can accomplish these movements using chemical messengers called **hormones** and various environmental stimuli. The most significant of these movements is seen in the way that plants can bend toward the sun to position themselves optimally for photosynthesis. This movement is called a **phototropism**. It is achieved using the hormone **auxin**, stored in the tip of the plant shoot and known for encouraging cell elongation and growth. A disproportionate increase in sunlight energy on one side of the stem encourages auxin to move toward the shaded side. There, the increased auxin causes cells and eventually the overall stem to elongate on that side only, producing a bend in the stem such that the plant orients itself toward the sunlight input. Several classes of plant hormones exist beyond auxin, each specializing in a different set of functions for the plant. Many are in fact involved in other movements like **gravitropism** (or geotropism) and **thigmotropism**, directed growth in response to gravity and touch, respectively. Gravitropism helps seeds germinate correctly such that the roots grow downward and the shoot upward. Thigmotropism allows for vines to wrap around the branches of other plants to prop themselves up closer to the sun.

EXERCISE
**16·1**

**Vocabulary Building.** *Explain the relationship between the following pairs of vocabulary terms.*

1. nonvascular plant, vascular plant

_____

_____

2. spore, seed

_____

_____

3. gymnosperm, angiosperm

_____

_____

EXERCISE
**16·2**

**Multiple Choice.** *Select the best response from the options provided to answer each question or to complete each statement.*

1. All members of kingdom Plantae are always
   a. autotrophic
   b. multicellular
   c. eukaryotic
   d. all of the above

2. Which of the following is *not* a characteristic of mosses?

    a. small size                       c. sporophyte-dominant

    b. nonvascular                  d. flagellated sperm

3. Typical plants should possess

    a. a cuticle                       c. vascular tissue

    b. flowers                        d. phloem

4. Flowers in angiosperms are analogous to what structures in gymnosperms?

    a. pollen                       c. seeds

    b. cones                      d. fruits

5. All of the following are associated with movement of water through the xylem *except*

    a. transpiration               c. active transport

    b. cohesion                 d. capillarity

6. The sporophyte generation of a plant produces spores; the gametophyte generation produces

    a. eggs                        c. seeds

    b. sperm                     d. both a and b

7. Which of the following characteristics is more typical of a monocot than a dicot?

    a. fibrous root system         c. netlike venation

    b. flower parts in multiples of four     d. two seed leaves

---

EXERCISE
## 16·3

**Short Answer.** *Write brief responses to the following.*

1. Name the significant evolutionary adaptation represented by each of the following plant subgroups, and then relate the structure to its function: mosses, ferns, gymnosperms, and angiosperms.

 

 

 

 

 

 

2. A woman has a houseplant situated in a brightly lit window. She has ensured that it is receiving the correct amount of light and water but has noticed that some of the leaves are showing yellow streaks. Provide a hypothesis to explain the problem.

_____

_____

_____

_____

3. Briefly describe *three* significant ecological roles of kingdom Plantae.

_____

_____

_____

_____

_____

_____

**EXERCISE 16·4**

**Labeling Diagrams.** *Fill in the blanks using the following terms to correctly label the diagram representing the alternation of generations life cycle.*

fertilization

germination

meiosis

mitosis

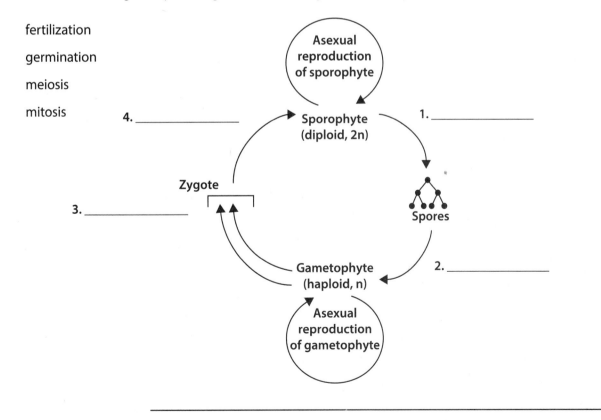

**Thinking Thematically.** *For each of the following themes of biology, choose a different concept from this chapter and explain how it provides a useful illustration of that theme.*

1. regulation and feedback

_____

_____

_____

_____

_____

_____

2. natural interdependence

_____

_____

_____

_____

_____

_____

3. continuity and change

_____

_____

_____

_____

_____

_____

## For Further Investigation

Take a walk with a notebook and pencil and/or a camera, and observe the plant world in your local ecosystem. Record any significant observations and document any characteristic structures. Attempt to find at least one representative for each of the following groups: mosses, ferns, conifers, monocots, and dicots.

# ANIMAL DIVERSITY
# AND MORPHOLOGY

# The Consumers

## Kingdom Animalia

Animals have established themselves as the dominant species in almost every habitat on Earth. From the great white shark that roams every ocean to the lion that rules the savannas of Africa, top animal predators are well known to us. We might live with or near domesticated animals like cats, birds, or horses. And let's not forget that we are animals ourselves!

Animal biodiversity extends far beyond the vertebrates just described. From the relatively simple sponge to the socially sophisticated bee and the intelligent squid, invertebrates truly dominate the globe. They account for approximately 95 percent of all animal species, are exceedingly diverse in habitat, feeding style, body structure, and behavior, and provide some essential ecosystem services for humans and other organisms.

## Animal Characteristics and Classification

To be considered a member of kingdom **Animalia**, an organism must possess several defining characteristics. First, animals (like fungi) are exclusively heterotrophic eukaryotes. Unlike fungi who absorb their already-digested food, animals are **ingestive heterotrophs**—they consume large, complex food and then digest it internally. Second, all animals are multicellular and rely on the process of **differentiation** to achieve cells specialized in structure and function. Effective communication and coordination between cells is essential, and that is facilitated structurally by the lack of a cell wall.

A member of the animal kingdom is also characteristically **motile** at some point in its life cycle. Although most animals familiar to us have obvious means of moving around in their various environments, some animals like barnacles are instead **sessile** as adults. These sessile animals typically possessed some form of motility during the larval stage. Whenever movement occurs in an animal, it is enabled by muscle and nervous tissue found uniquely within the kingdom.

The vast majority of animals are characterized by some form of sexual reproduction as well. Although the specific mechanisms vary markedly among different species, it always involves at the most basic level the fertilization of an egg by a sperm to establish the zygote. The zygote, through successive, well-orchestrated mitosis and cytokinesis events, creates an embryo. Often, an animal will go through discrete stages of development, called **indirect development**. The caterpillar that morphs into a butterfly clearly exemplifies this pattern. When domestic cats and dogs develop, they instead demonstrate **direct development**. Although there are still observable points of contrast between the juvenile and adult stages, the juvenile for the most part resembles a smaller version of the adult.

The animal kingdom is subdivided into two large, practical groups previously mentioned, the **invertebrates** and the **vertebrates**. While the vertebrates share a relatively close evolutionary history and thus numerous homologous traits, the invertebrates really only have in common what they lack—a backbone, or **vertebral column**, that provides protection for the central nerve cord in vertebrate animals.

From a phylogenetic perspective, animals are subdivided into more than thirty phyla, but only one phylum contains all vertebrate species. The other phyla are made of several groups of worms, many deep-sea creatures, arthropods, and mollusks, to name a few. Classification of animals into phyla depends upon the early expression of several developmental genes. These include genes involved in the development of true tissues, of discrete embryonic tissue layers, of any symmetry observed within the body plan, and of the developmental pattern of the digestive tube.

## Invertebrate Biodiversity

The most ancestral and also the most structurally simple animal group is phylum Porifera, the sponges (refer to Figure 17.1a). Most are marine, although there are notable freshwater varieties as well. Sponges are unique among the animal kingdom for several reasons, including their **asymmetry** and lack of true tissues. Sponges certainly are multicellular and do possess specialized cells, but those cells do not group together into collectively functioning tissues within the organism. Instead, specialized cells like **choanocytes** are scattered throughout the sponge body. These flagella-lined cells surrounding pores in the sponge body provide a constant intake for water and the small organic matter that the sponge can filter out, deliver to its cells, and call food. In fact, these choanocytes closely resemble the animal-like protist from which the entire animal kingdom likely evolved.

**Figure 17.1a** A Sponge

Another ancient group of invertebrate animals is phylum Cnidaria, including the sea jellies, sea anemones, and corals (refer to Figure 17.1b). Like the sponges, cnidarians are predominantly marine with relatively few freshwater representatives. Unlike the sponges, however, these organisms possess true tissues and a body plan with **radial symmetry**, characterized by a central region with appendages branching out from there. Known for their specialized stinging cells called **cnidocytes**, cnidarians are predators. The sea jellies, which are motile as adults, are described as

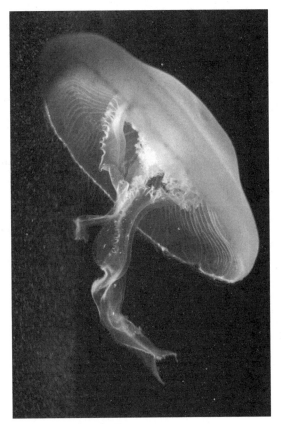

**Figure 17.1b**  A Sea Jelly

being in the **medusa** phase when they have reached maturity. This is sharply contrasted by the life cycles of the corals and anemones, which have a motile larval stage but remain sessile as adults in the **polyp** phase.

As major representatives of invertebrates, worms are so plentiful and diverse that they make up close to ten animal phyla on their own. There are flatworms (phylum Platyhelminthes), roundworms (phylum Nematoda), ribbon worms (phylum Nemertea), and segmented worms (phylum Annelida). Regardless of the terrestrial earthworms that humans are familiar with (see Figure 17.2a), most free-living worm phyla are marine. The others are animal parasites that spend a portion of their life cycle in a freshwater habitat, likely one that the animal lived in or drank from.

Worms are characteristically elongated, **bilaterally symmetrical** animals with no appendages and few sensory structures. Many have a relatively complex internal anatomy unmatched by their comparatively simple exterior. For example, earthworms and other segmented worms possess a highly specialized digestive tract and an efficient, closed circulatory system. They also are **hermaphroditic**, possessing both male and female reproductive organs at maturity. That feature allows them to double up on their reproductive success with each mating event, helpful especially when you live in a challenging habitat to find a mate, like the compact soil.

Mollusks, as seen in Figure 17.2c, include a variety of bilaterally symmetrical invertebrates like snails and slugs (class Gastropoda), squid and octopus (class Cephalopoda), and oysters and clams (class Bivalvia). Members of phylum Mollusca display increased complexity in form from most worms, as they are true **coelomates**. These are animals that possess a body cavity completely derived from the middle tissue layer during embryonic development. The result is separate, lined body cavities that allow for optimal skeletal muscle movement and internal digestive and circulatory function. The internal organs of a mollusk are concentrated in an area called the **visceral mass**, and in most species the epidermis secretes a tough outer shell called a **mantle**.

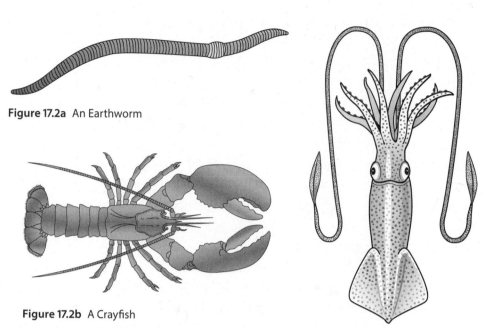

**Figure 17.2a** An Earthworm

**Figure 17.2b** A Crayfish

**Figure 17.2c** A Squid

Another significant group of bilateral invertebrate coelomates is the arthropods, so named for their highly adapted, jointed extensions of the body called **appendages**. Members of phylum Arthropoda comprise a huge and diverse group that constitutes approximately 75 percent of all invertebrate species (see Figure 17.2b). They include insects (class Insecta), spiders and ticks (class Arachnida), and lobsters and barnacles (subphylum Crustacea), successfully occupying nearly every habitat on Earth. All arthropods possess an **exoskeleton** composed of the carbohydrate chitin for protection and support and **compound eyes** for reception of visual stimuli. Depending on the type of arthropod, various specialized appendages, like antennae, mouthparts, and walking legs or wings, provide adaptive advantages for the organism in its environment.

The invertebrate phylum most closely related to the vertebrates is Echinodermata. The echinoderms, another coelomate group, are characterized by radial symmetry as adults, a trait that fits their bottom-feeder lifestyle. Their larval form of echinoderms is instead bilateral. This group includes marine organisms like sea cucumbers, sand dollars, and the characteristic sea stars. They have a specialized **water vascular system**, a network of water-filled canals that permits movement of the numerous, sucker-like **tube feet**.

# Vertebrate Characteristics

Vertebrates are united through their possession of several characteristics beyond the obvious vertebral column. The vertebral column is there to protect the **dorsal nerve cord** within, a feature that allows the complex brain to communicate with the rest of the body. Early in embryonic development, the vertebral column actually existed as a flexible **notochord** instead, a feature retained by some invertebrate chordates. Similarly, **pharyngeal gill slits** are present during embryonic development but are only retained as functional gills by the aquatic vertebrates.

The **endoskeleton**, or internal skeleton, typical of members of subphylum Vertebrata, makes them distinct from the invertebrate chordates. The endoskeleton is composed of either cartilage or bone, depending on the class. Although the exact composition of the skeleton varies greatly, it always includes a **cranium**, a protective covering for the complex brain, as well as the **vertebrae,** individual segments within the vertebral column already discussed.

# Fishes

Fishes represent a large and diverse group of vertebrates. Unifying the group are their paired fins and streamlined bodies that make them very successful at aquatic life. Mucous secretions from the integument help reduce friction in the water further. Most fishes have jaws and an integument composed of scales, although they vary in type. A system of canals within the skin called the **lateral line** enable fishes to detect subtle vibrations in the water.

Fishes are also **ectotherms**, possessing a body temperature dictated by the temperature of the surrounding water. Internally, the circulatory system of fishes is propelled by the pumping action of a two-chambered heart comprising an upper and lower chamber. This type of two-chambered design is indicative of a single circuit for the bloodstream that both supplies the body with oxygen and absorbs new dissolved oxygen from the water using the gills. The simple circuit permits mixing of oxygen-rich and oxygen-poor blood to a certain degree and thus is less efficient than the double circuit design that later evolved in other lineages.

One major class of fishes (class Chondrichthyes) includes all cartilaginous fishes like sharks, rays, and skates (see Figure 17.3a). Most are characterized by internal fertilization and development and are **viviparous**, giving birth to live young. A unique adaptation called the **ampullae of Lorenzini** allows the cartilaginous fishes to detect slight electrical impulses in the water emitted from the nervous and muscular system of potential prey.

Bony fishes (superclass Osteichthyes) include two major classes, the ray-finned and lobe-finned fishes (see Figure 17.3b). While the **ray-finned fishes** have only thin bones making up the fin structure, **lobe-finned fishes** possess soft tissue at the base of the fin as well. As opposed to the cartilaginous fishes, bony fishes typically engage in external fertilization in a **spawning** event, when massive quantities of gametes are released, and they have also become very well adapted to both marine and freshwater ecosystems. A unique adaptation possessed by the bony fishes is the **swim bladder**, a gas-filled internal float that allows the organisms to physiologically adjust their buoyancy. Another adaptation is the **countercurrent flow** design observed in their gills. The oppositional flow of deoxygenated blood in blood vessels and water flow over the gills maximizes the efficiency of gas exchange and thus benefits the organisms metabolically.

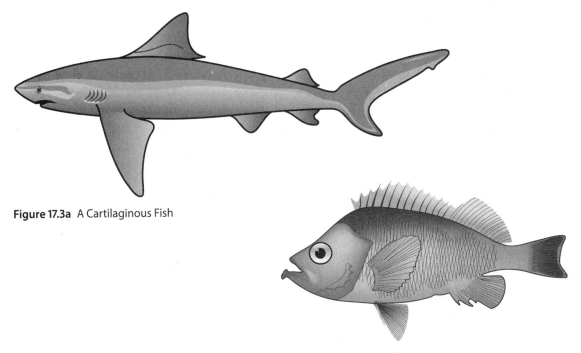

**Figure 17.3a** A Cartilaginous Fish

**Figure 17.3b** A Bony Fish

# Amphibians

Amphibians, somewhat analogous to the bryophytes of the plant kingdom, are terrestrial but remain dependent on aqueous ecosystems to varying degrees. While adults possess lungs as a respiratory organ, they are relatively small and are supplemented by the action of the moist skin. Like their fish ancestors, amphibians are ectothermic. The groups differ in that amphibians evolved limbs to replace fins and utilize a three-chambered heart to drive the circulatory system. Composed of two upper atria and one lower ventricle, the amphibian heart allows for a **systemic circuit** to supply oxygenated blood to the body and a separate **pulmocutaneous circuit** to resupply deoxygenated blood at the lungs and skin. More efficient than the two-chambered approach, the single ventricle of the three-chambered heart still allows for mixing of oxygen-rich and oxygen-poor blood.

Members of the amphibian class reproduce through external fertilization through the coordinated release of gametes from both sexes. A marked form of indirect development called **metamorphosis** is characteristic of this group and exemplified by the tadpole that matures into a frog (see Figure 17.4).

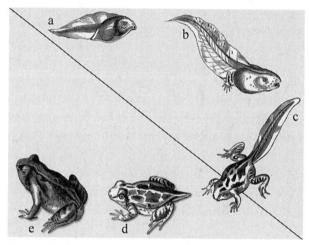

**Figure 17.4** Metamorphosis in Amphibians

# Reptiles and Birds

Once animals were well established in the terrestrial world, a new group evolved that eventually dominated the planet for millions of years. Whereas their amphibian ancestors toed the line between water and land, reptiles fully committed to terrestrial life (see Figure 17.5a). They evolved adaptations such as scales made of **keratin**, a waterproof protein, and abandoned any use of respiratory skin for fully developed, more efficient lungs. A major terrestrial adaptation observed in reptiles is the **amniotic egg**, a complex structure that protects the developing embryo from the harsh external environment while at the same time supplying it with nutrients and separating wastes.

Reptiles demonstrate interesting variation in heart structure. Some have three chambers like amphibians, while others have three chambers with a partial **septum**, or wall, within the ventricle region. Only the crocodiles have a complete septum and thus a four-chambered heart.

Nonavian reptiles are ectothermic like amphibians, and although the amphibian and reptilian brain are about equal in size, reptiles have an enlarged cerebrum that enables more complex behavior. Reptilians display a variety of means of giving rise to young; they can be **oviparous** (producing eggs that hatch externally), **ovoviviparous** (producing eggs that hatch internally), and viviparous.

**Figure 17.5a** A Reptile (a Crocodile)

**Figure 17.5b** An Avian Reptile (a Bird)

It surprises many people to learn that birds have recently been reclassified as avian reptiles (see Figure 17.5b). Because of high energy requirements of flight, birds evolved several traits that make them quite distinct from their reptilian ancestors. Feathers made from modified scales and hollow bones evolved to facilitate flight. Given the high energy requirements of that characteristic activity, a four-chambered heart and specialized flow-through lungs with **air sacs** also evolved to help maximize metabolic efficiency. Birds are **endothermic**, controlling their internal body temperature physiologically.

## Mammals

When reptiles had reached their dominance as myriad dinosaur species, mammals were present but generally restricted to small, burrowing creatures. The catastrophic event that led to the mass extinction—approximately 70 percent of all life on Earth at the time went extinct—is indicated by geologic data to be either an eruption of a super volcano or an impact of a massive meteorite. The habitat of the small, burrowing mammals proved to be protective from the destructive atmosphere that certainly followed the catastrophic event for some time, and the survivors of the mass extinction eventually flourished. Extensive divergence of the relatively few mammal species at the time led to the thousands of mammal species that thrive today.

Mammals, named for the **mammary glands** that produce and secrete milk for the young, are practically subdivided into three groups based on the manner in which the offspring are introduced into the world (see Figure 17.6). A small and non-representative group, the **monotremes** include only two species of echidna and one species of duck-billed platypus. They hatch their young from leathery, reptile-like eggs after internal fertilization takes place. Their milk oozes from patches on the skin instead of through nipples.

A more recognizable group of mammals, the **marsupials**, display vivipary, the ability to give birth to live young following a period of internal fertilization of the zygote and development of the embryo. Marsupials differ from the final group called placental mammals because they spend very little time in the mother's womb before crawling out and into the pouch she carries on the ventral side of her abdomen. Inside this pouch, the embryo continues development while having access to milk from mom's mammary glands.

**Placental mammals**, the group to which we belong, are characterized by a much longer period of internal embryonic development within the mother's **placenta**. Once born, the young mammal then has access to the mother's mammary glands for consumption of nutritious and immunity-boosting milk.

**Figure 17.6** Three Types of Mammals

Create a caption for Figure 17.6 to illustrate the key differences between three three types of mammals. What is observed and what can instead be inferred?

Regardless of the group to which they belong, all mammals possess several unifying characteristics beyond the mammary glands. These include endothermy, internal fertilization, diet-specific specialized teeth, a very efficient four-chambered heart, and highly developed brains and nervous systems. The frontal lobe of the brain called the **cerebrum** is most highly developed in mammals, enabling complex behaviors and levels of intellect.

**Vocabulary Building.** *Explain the relationship between the following pairs of vocabulary terms. Provide an example of an animal or a group of animals with each characteristic.*

1. sessile, motile

   _____

   _____

2. indirect development, direct development

   _____

   _____

3. invertebrate, vertebrate

   _____

   _____

4. oviparous, viviparous

   _____

   _____

5. endotherm, ectotherm

   _____

   _____

**Multiple Choice.** *Select the best response from the options provided to answer each question or to complete each statement.*

1. All members of kingdom Animalia are always
   - a. prokaryotic
   - b. colonial
   - c. ingestive heterotrophs
   - d. vertebrates

2. All of the following are involved in sexual reproduction and development in animals *except*
   - a. a coelom
   - b. hermaphroditism
   - c. amniotic egg
   - d. placenta

3. Specialized stinging cells in a sea jelly are called
   - a. septa
   - b. choanocytes
   - c. cnidocytes
   - d. tube feet

4. An example of a radially symmetrical invertebrate is a
   a. sea star
   b. lobster
   c. sponge
   d. segmented worm

5. Of the following, the vertebrate group with the fewest number of chambers in the heart is
   a. Osteichthyes
   b. Amphibia
   c. Aves
   d. Reptilia

6. All of the following are examples of vertebrates *except*
   a. cartilaginous fishes
   b. echinoderms
   c. marsupials
   d. avian reptiles

7. The only oviparous mammals are the
   a. placentals
   b. reptiles
   c. marsupials
   d. monotremes

EXERCISE
17·3

**Short Answer.** *Write brief responses to the following.*

1. Describe the major characteristics that distinguish a vertebrate from invertebrate animals. What is the significance of some of these characteristics?

   _____

   _____

   _____

   _____

   _____

   _____

2. Which non-phylogenetic group of animals seems to be more successful on Earth, those with radial symmetry or those with bilateral symmetry? Explain.

   _____

   _____

   _____

   _____

   _____

   _____

   _____

3. Compare and contrast the cartilaginous fishes from the bony fishes.

_____

_____

_____

_____

_____

_____

**Interpreting Diagrams.** *Examine the following diagram representing a phylogenetic tree of vertebrates. Use the information in the diagram to answer the questions that follow.*

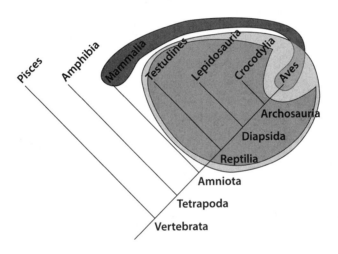

1. Which group diverged more recently, Testudines or Lepidosauria?

_____

2. Which group is least closely related to the rest?

_____

3. Although mammals and birds are not as closely related as some of the groups shown, they are similar in that they independently evolved what important trait?

_____

**Thinking Thematically.** *For each of the following themes of biology, choose a different concept from this chapter and explain how it provides a useful illustration of that theme.*

1. energy and organization

_____

_____

_____

_____

_____

_____

2. form facilitates function

_____

_____

_____

_____

_____

_____

3. continuity and change

_____

_____

_____

_____

_____

_____

_____

## For Further Investigation

Use the Internet to investigate the box jelly, a cnidarian native to the coasts around Australia. Discover what distinguishes this tiny sea jelly from its other jelly relatives. What do these physiological or behavioral differences suggest about the evolution of the box jelly?

# The Body Systems

## Human Physiology

From an evolutionary perspective, modern humans (*Homo sapiens sapiens*) are the new kids on the block. What this means, however, is that we have inherited the evolutionary "wisdom" of our many ancestor species. The result is an exceedingly complex organism composed of eleven integrated organ systems, each intricately orchestrated and regulated. A brief introduction to the structure and function of the human body's systems is presented here.

## Our Fundamental Body Plan

Four fundamental tissue types exist in humans. **Muscle tissue** is classified as either skeletal, smooth, or cardiac based on the structure and function of the tissue. **Skeletal muscle** controls voluntary movement of the skeleton when it contracts, or shortens. Skiing, typing, and driving are activities all requiring coordination of various skeletal muscles to perform. **Smooth muscle** is instead responsible for most involuntary movements, like your stomach churning in an effort to digest a meal. **Cardiac muscle** is an essential tissue restricted to the heart. It is involuntary in that it is not under conscious control of the brain but differs from smooth muscle in that it generates its own contraction independent of the brain. Cardiac muscle is also unique in that it is designed to never stop contracting as long as the human is alive.

Nervous tissue is comprised predominantly of **neurons**, elongated nerve cells capable of transmitting an electrochemical message. When muscle tissue contracts, it does so upon the signaling from a nerve. **Connective tissue** is made up of cartilaginous cells embedded in an extracellular matrix and generally provides cushioning, support, and integration for the body. Connective tissue varies greatly in function depending upon its structure. At one extreme is bone, where mineralization of the cells is present, and at the other extreme is blood, cells scattered in an aqueous medium. Somewhere in the middle of this spectrum is cartilage, found at all joints of the body to prevent bone-to-bone contact and provide shock absorption. The last major type is **epithelial tissue**, which consists of tightly bound cells that form a protective lining to cover major organs and the entire body itself. Depending on the location and function of the epithelial tissue, its arrangement of cells may be simple (one layer), stratified (multiple layers), or pseudo-stratified (one layer with irregularly shaped cells that appear to be in multiple layers). The individual cells themselves may be squamous (flattened), columnar (elongated), or cuboidal (rounded).

Using these four fundamental tissues, humans are able to create all of the **organs** that then work together within an organ system to achieve a particular goal for the collective body. Most organ systems are housed in various cavities of the body. The **cranial cavity** contains the brain, and the **spinal cavity** the spinal cord, collectively containing the central nervous system. The **thoracic cavity** is the location of the respiratory system and is literally and figuratively the heart of the circulatory system. Separating the thoracic cavity from the abdominal cavity below is a thin sheet of muscle called the **diaphragm**, which directs inhalation and exhalation of air. The **abdominal cavity** houses the digestive system and some excretory organs, while the **pelvic cavity** provides a location for the reproductive system and the remainder of the excretory system.

The remaining systems not yet mentioned (integumentary, endocrine, immune, skeletal, and muscular) are not located within a cavity but are instead spread throughout the body, helping to protect, coordinate, and integrate all of the various structural components into one functional organism.

## Standing Tall and Getting Around: The Muscular and Skeletal Systems

The muscular system includes all muscular tissue that can contract and relax to provide movement within the body and of the body itself. Skeletal muscle is the most obvious, lying just underneath the skin and a layer of fat. Skeletal muscle is arranged into increasingly smaller bundles until reduced to the **filament** level, at which actual contraction occurs. A muscle like the bicep is composed of smaller bundles called **fascicles** bound together by connective tissue. These fascicles in turn are made up of bundles of individual muscle cells or **fibers**, elongate and multinucleate.

Within the muscle cells are densely packed, threadlike proteins called **myofibrils**. These myofibrils are composed of two types of proteins, the thick filament **myosin** and the thin filament **actin**, arranged into the structural unit of muscle contraction, the **sarcomere**. When the actin is pulled over the myosin from opposite directions toward the midline of the sarcomere, the overall muscle shortens, or contracts. When the myosin releases the actin and it slides back to the starting position, the length of the sarcomere increases, and the muscle relaxes. This mechanism for muscle function requires ATP and is referred to as the **sliding-filament theory**.

Recall that smooth muscle is responsible for involuntary internal movements like moving food through the digestive tract and blood through the arteries. Its component cells are structurally different from skeletal muscle cells; they are not multinucleate, they are arranged into sheets, and the matrix of their surrounding connective tissue is loosely configured. Cardiac muscle cells have properties of both other muscle types; they are striated with filaments like skeletal muscle but operate involuntarily and possess just one nucleus per cell.

In order for skeletal muscles to allow movement of the body, they must be attached to the skeletal system. **Tendons** are composed of fibrous connective tissue and attach skeletal muscle to bone at regions where two or more bones meet, called **joints**. Bones are then connected to each other by another type of fibrous connective tissue called **ligaments**. Usually, yet another type of connective tissue called **cartilage** provides cushioning and shock absorption within and surrounding each joint. The skeleton itself is actually divided into two subparts: the axial skeleton and the appendicular skeleton. Together the skull, vertebral column, and the ribs comprise the **axial skeleton**, built for protection of the most essential organs like the brain, spinal cord, heart, and lungs. The **appendicular skeleton** instead provides the framework for the appendages of the body, the arms and legs (see Figure 18.1).

The adult human body contains a total of 206 bones; more are present in the fetus for increased flexibility during birth but eventually fuse during infancy in a process called **ossification**. Most people think of bone as primarily mineral, but in fact bone is a living tissue ripe with blood vessels and active cells. Recall that bone is a type of mineralized connective tissue, giving it its unique durable form. The mineral salts that make up bone serve as a calcium and phosphorous reserve for the body. The majority of bone is called **compact bone** and is made up of individual units called **Haversian canals**. Through the central canal, the blood travels, supplying oxygen and nutrients to the bone tissue. Bone cells known as **osteocytes** deposit new mineral

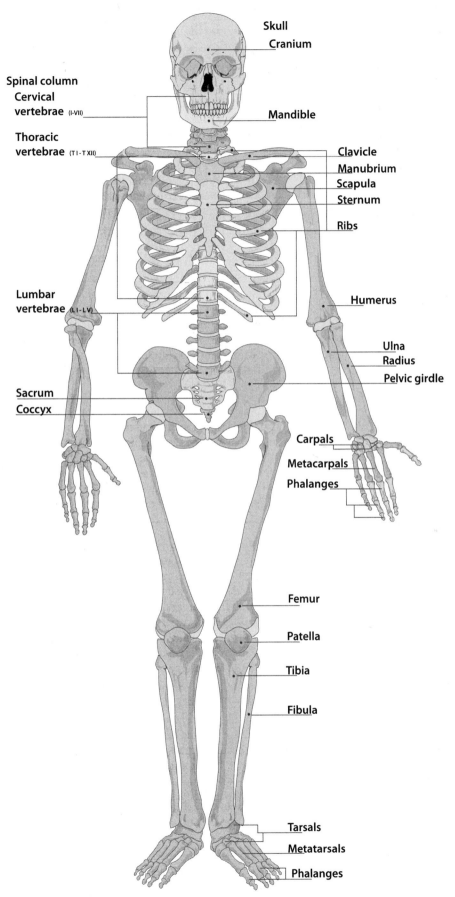

**Figure 18.1** The Human Skeletal System

layers in concentric circles around the central canal. Some bone has regions characterized as spongy instead of compact. **Spongy bone**, while porous in nature, is actually very strong and is designed to divert the stress placed on one point of impact over a larger surface area. A soft tissue called **bone marrow** is also present in many bones. The more familiar red marrow produces blood cells and platelets and thus contains many stem cells; the less familiar yellow marrow contains fat cells that supply long-term energy to long bones like those of the arms and legs.

## Protection and Fighting Back: The Integumentary and Immune Systems

The **integumentary system** includes almost everything you see of your body from the outside, namely skin, hair, and nails. All three of these structures are composed of the protein **keratin** arranged in slightly different ways. Keratin is waxlike, thus providing watertight skin except at pores, where fluid can be released in a controlled fashion. The largest organ of the body, skin is made up of two discrete layers, the epidermis and the dermis (see Figure 18.2). The upper **epidermis** is regenerated constantly as skin cells from the top of the dermis die and become increasingly flattened as they are pushed toward the surface. As dead cells at the surface experience abrasion, they fall off and make room for lower cells.

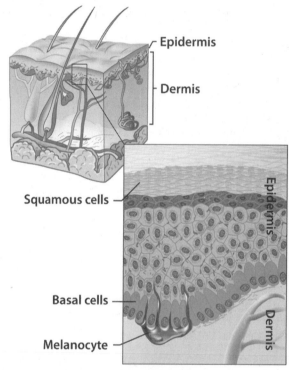

**Figure 18.2** The Skin

The **dermis** is the site of living cells in the skin. It is also the location of nerves to provide sensation and blood vessels to provide oxygen and other nutrients. Two types of **exocrine glands**, those that secrete a substance directly onto the place it is designed to work, are also found within the boundaries of the dermis. **Sweat glands** help the body maintain thermal homeostasis in a process called **evaporative cooling** (skin is also involved in the shivering response when the body is too cold, for tiny muscles under the surface of the skin tweak individual hairs and lead to the formation of goose bumps). **Sebaceous glands** release oily **sebum** as a natural conditioner for the skin, hair, and nails. Specialized cells called **melanocytes** produce the dark pigment **melanin**; the higher the concentration of melanin in the skin, the darker the color of the skin overall, and the more protective benefit gained from UV light.

The skin not only protects all of the internal organs from injury and the elements, but it also acts as the first line of defense for the **immune system**. Preventing a plethora of pathogens like viruses and bacteria from entering the body is a nonspecific mechanism of defending the body from danger. Where there are natural openings to the skin, like in the ears, nose, and throat, there are layers of cilia or mucus present. These provide backup protection by trapping particulate matter before it makes its way farther into the body. Other potential sites of infection come equipped with a natural antibiotic substance, like the eyes with their tears.

If something potentially hazardous gets past the skin and its allies, then the second line of defense is called into action. This involves the **inflammatory response**, a rush of blood to the infected or injured area in response to a release of the protein **histamine**. With blood comes the swelling, redness, and heat characteristic of an inflammation, but more importantly it carries all of the essential ingredients necessary to try to solve the problem at hand. In any case, energy will be needed for the active repair of damage to the body's tissues, so red blood cells, or **erythrocytes**, are helpful in delivering oxygen for cell respiration and ATP production. If there is an injury to blood vessels, then **platelets** and blood proteins are needed to form a clot and stop the leaking vessel (see Figure 18.3).

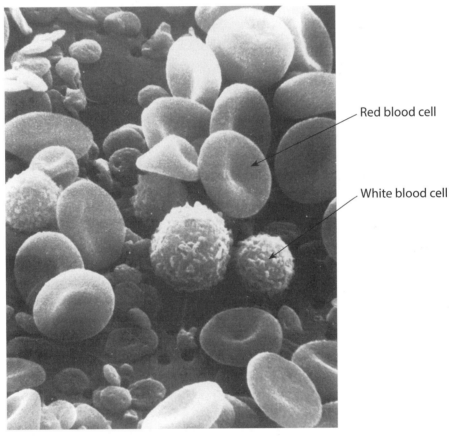

Red blood cell

White blood cell

**Figure 18.3** Blood Cells

If instead an actual infection is present, then white blood cells, or **leukocytes**, specifically seek out and destroy the pathogen in what is recognized as the third line of defense. **B leukocytes** are utilized for their ability to secrete **antibodies**, Y-shaped proteins that recognize the specific pathogen, bind to it, and tag the invader for destruction by other cells. The counterpart **T leukocytes** patrol the body looking for its own cells that have already been infected by the pathogen. Infected cells are recognizable because they display a piece of the virus or bacterium on their outer surface like a distress signal. The T leukocyte then knows to destroy the infected cell, thereby preventing it from potentially releasing more pathogen into the body.

# All About Oxygen: The Circulatory and Respiratory Systems

The aforementioned blood needs a way to constantly move throughout the body, and this role is filled by the organs of the **circulatory system**. Often also called the *cardiovascular system* in humans, the circulatory system is composed of the heart, blood vessels, and the blood they contain (see Figure 18.4). The human **heart** is typical of any mammal, a muscular, four-chambered pump with a complete septum that separates the right and left sides. The two upper chambers, or **atria**, are separated from their lower compartments, the **ventricles**, by muscular valves that open and close to direct blood flow. The heart maintains its own electrical impulse independent of the brain in a region in the right atrium called the **pacemaker**. The impulse then runs throughout the heart tissue in a well-orchestrated wave.

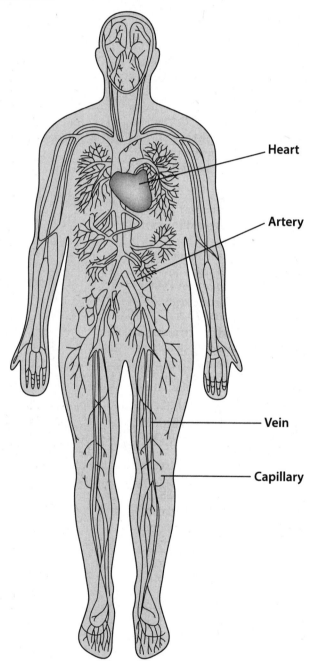

**Figure 18.4**  The Circulatory System

Create a caption to Figure 18.4 to relate the structure of the circulatory vessels to their specific transport function.

Blood that first entered an atrium is pushed into the corresponding lower ventricle, and then it is forced out of the ventricle and carried out of the heart altogether by means of a blood vessel. Because the vessel is moving blood away from the heart, the vessel is called an **artery**. The thick, muscular walls of arteries allow them to withstand the force of the heart; the force of the blood on the walls of the arteries is the means by which **blood pressure** is measured. As arteries move farther away from the heart, they continually branch into smaller vessels and eventually are termed **arterioles**. The branching pattern continues from there, eventually creating vessels whose walls are only one-cell thick. These tiny vessels known as **capillaries** are the only tubes that can exchange materials between the blood and the surrounding tissues. Oxygen and nutrients can be exchanged for carbon dioxide and other metabolic waste products where these vessels are concentrated in a structure called a **capillary bed**. As blood moves into larger tubes again on the other side of the capillary bed, these tubes are carrying blood back toward the heart. The initial smaller vessels are called **venuoles**, and the eventual larger vessels are termed **veins**. The largest vein in the body, the **vena cava**, allows blood to finally reach the heart again to complete the circuit.

The heart has two completely distinct sides in order to completely separate oxygen-rich from oxygen-poor blood and thus to maximize efficiency. The right atrium receives the **deoxygenated** blood from the body, devoid of oxygen after delivering its contents to the body's cells. It moves the blood into the right ventricle and then out of the heart through the **pulmonary arteries**. These tubes carry the blood directly to the lungs in what is known as the **pulmonary circuit**. At the lungs, a new supply of oxygen is picked up by the blood, and then this freshly **oxygenated** blood is carried back to the heart through the **pulmonary veins**. The oxygen-rich blood moves first into the left atrium, then into the left ventricle, and finally out of the heart through the **aorta**, the largest artery of the body. The aorta represents the beginning of the **systemic circuit**, the one that delivers the blood to the body again.

The capillary bed that surrounds the lungs represents the integration of the circulatory system and the **respiratory system**. Consisting of the oral and nasal cavities, the pharynx, trachea, and lungs, the respiratory system is designed to pull air in and out of the lungs for gas exchange (see Figure 18.5). Air enters the body through the **oral** and **nasal cavities** and is pulled through

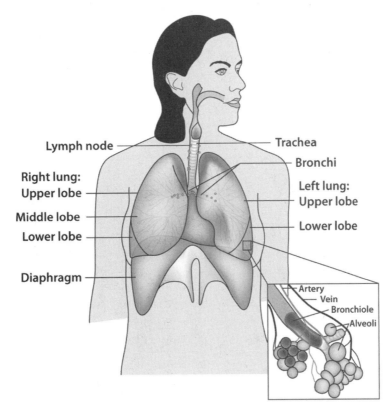

**Figure 18.5** The Respiratory System

the **pharynx** (throat) and into the trachea. The **trachea**, or windpipe, is a tube ringed with tough cartilage that ensures that the trachea does not collapse and prevent air from reaching the lungs. The trachea eventually splits into two smaller tubes, the bronchi.

One **bronchus** moves air to the right lung, the other bronchus to the left lung. Each bronchus further branches into smaller tubes called **bronchioles**, and these bronchioles deliver the air to the terminal air sacs, called *alveoli*. Each **alveolus**, which appears like a bundle of grapes, has increased surface area for maximum gas exchange. The alveolus is surrounded by a capillary bed, and the gradient is such that oxygen diffuses from the air in the lungs to the blood in the capillaries, where it is then picked up by the hemoglobin in a red blood cell. Carbon dioxide follows the opposite path: gas is exchanged, and the blood is revived with energy-providing oxygen for cell respiration.

Inhalation of air into the lungs occurs by means of two types of muscle: the thin **diaphragm** that sits just below the lungs and above the stomach and the **intercostal muscles** that connect the ribs to each other. As the diaphragm and the intercostal muscles contract, a negative air pressure is created in the space inside the lungs, and air is pulled into the lungs via nostrils or an open mouth.

## Taking in Nutrients and Sorting out Wastes: The Digestive and Excretory Systems

Cells don't rely only on the respiratory system for oxygen; they also need extracted nutrients from the food we eat and the action of the **digestive system** (see Figure 18.6). Humans take in the food we eat through the **mouth**, the site of ingestion. Here, all food starts **mechanical digestion** as the teeth and tongue physically tear the food into smaller pieces. **Chemical digestion**, the breaking down of polymers into monomers using enzymes, also begins as the **salivary amylase** in saliva gets to work on the starches in our food. As the tongue pushes food toward the pharynx, the swallowing reflex is triggered, and the food remains, called a **bolus**, are forced into the **esophagus**. This long tube runs just behind the trachea, so with swallowing also comes a flap of tissue called the **epiglottis** that quickly acts to cover the usually exposed trachea until food has safely moved

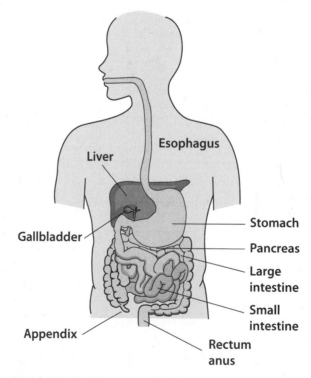

**Figure 18.6** The Digestive System

into the correct tube. Smooth muscle creates wavelike contractions of the esophagus called **peristalsis**, moving food down to the J-shaped **stomach**.

The stomach is the next stop for digestion; its muscular churning action contributes to additional mechanical digestion, while the secretion of acid into the stomach activates the enzyme pepsin. **Pepsin** acts to digest polypeptides into smaller units, so the stomach is the center of protein digestion. From the stomach, the soupy **chyme** moves into the **small intestine**, a very long yet narrow tube. The first third of the small intestine focuses on lipid digestion, as **bile** is secreted into the tube from the **liver**, where it was made, and the **gallbladder**, where it was stored. Bile emulsifies fats, breaking them down into small fat globules. The remaining two-thirds of the small intestine is the site of absorption of all of the nutrients that have been digested thus far. Its lining has folds with even smaller folds on top of them; these **villi** and **microvilli** allow for maximum absorption of nutrients from the interior space of the small intestine and into the capillary beds that surround the tube.

After the nutrients from digested food are absorbed into the body, whatever is left at the end of the small intestine is considered digestive waste. This must be eliminated from the lumen of the intestinal tract, so wavelike contractions continue to push the waste into the **large intestine**, or **colon**. Shorter in length but larger in diameter than the small intestine, the colon is structured for the reabsorption of water into the bloodstream, which results in the solidification of wastes into feces. Temporarily stored in the **rectum**, feces are finally eliminated from the body through the **anus**.

As important as is delivering nutrients to all of the body's cells is excreting metabolic wastes that the same cells have produced. Metabolic wastes are distinct from digestive wastes just described, as metabolic wastes result from the chemical activity of the body's cells and rely on the actions of the bloodstream and the paired **kidneys** for actual excretion (see Figure 18.7).

Each kidney is composed of more than one million **nephrons**, the individual functional unit of the kidney. A small tuft of capillaries delivers blood to each nephron and forces any non-cellular, non-macromolecular contents into the surrounding space of the nephron. From there, the filtrate is moved through a long, winding tubule that intertwines with a capillary bed to allow for maximum exchange between the kidney and the bloodstream. Anything the body wants to get rid of remains within the kidney tubule, while anything that is suitable for being retained by the body is absorbed back into the blood.

Each nephron contributes its final excretory product, now considered **urine**, to the **collecting duct**, where it gathers with the other filtrate from that kidney and is moved via the tubular

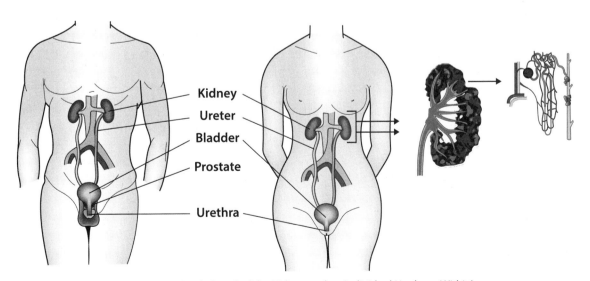

**Figure 18.7** The Excretory System (with Detail of the Kidney and an Individual Nephron Within)

**ureter** to the bladder, where it is stored. Finally, yet another tube called the **urethra** moves the urine out of the body.

Although the kidneys are definitively the primary organ of excretion in humans, the skin and the lungs play supplementary roles. The skin, through sweat, also releases nitrogenous wastes in relatively larger volumes of water than the kidneys do through urine. The lungs are involved not with the excretion of nitrogenous wastes but instead with that of the carbon dioxide wastes from cellular respiration.

## Communication and Regulation: The Nervous and Endocrine Systems

Keeping all of the other systems coordinated and in homeostasis are the nervous and endocrine systems. The nervous system is organized into two subparts, the **central nervous system (CNS)** and the **peripheral nervous system (PNS)**. The CNS is composed of the brain and spinal cord and functions in overall control and coordination, while the PNS consists of all of the body's nerves, the majority of which branch out from the spinal cord and span the entirety of the body (see Figure 18.8). The PNS has significant subdivisions as well, namely the **somatic nervous system** and the **autonomic nervous system**. Somatic nerves initiate voluntary actions that are under conscious control, like the ones responsible for initiating contraction of skeletal muscle. Auto-

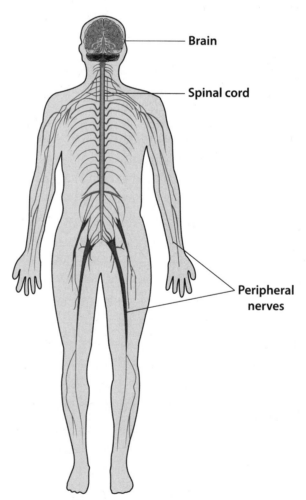

**Figure 18.8** The Nervous System

nomic nerves are controlled involuntarily by the body and initiate actions within the internal organs. When contractions are initiated within the digestive tube or the blood flow is increased to the lungs, the autonomic system is at work. The autonomic nervous system itself is finally subdivided into the **sympathetic nervous system**, that which initiates the "fight-or-flight" response when faced with a threat, and the **parasympathetic nervous system**, which is responsible for returning the body to normal after the excitatory "fight-or-flight" response.

The neurons that comprise the CNS are called **interneurons**. The brain, for example, contains a vast network of branched, connected interneurons. These many connections between cells are the basis of learning. As a specific set of interneurons communicate repeatedly with one another, the signals between them become stronger, and the stored information can be accessed more readily—thus the brain has learned. Weak connections can also be broken in a pruning process to improve efficiency of brain function. The neurons of the PNS are considered **sensory neurons**, those that perceive stimuli from the environment and direct the message to the brain, or **motor neurons**, those that travel from the CNS to the PNS in order to carry out an appropriate response. Often motor neurons initiate muscle contraction or secretion of a hormone from a gland.

Neuron structure is quite specialized to carry out the unique function of transmitting an electrochemical signal (see Figure 18.9). Tentacle-like projections called **dendrites** receive information from a sensory organ or another neuron and direct the signal to the **cell body**. There, the

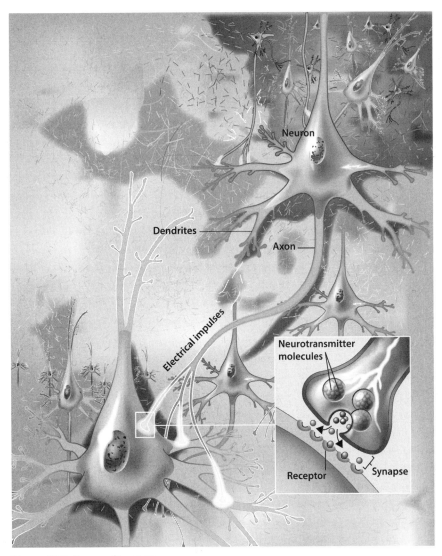

**Figure 18.9** Neuron Structure

nucleus of the cell remains along with the majority of the cytoplasm. The signal continues down the thin, elongated **axon** until it reaches the **axon terminal**. The majority of the length of the axon is wrapped in a **myelin sheath** as insulation, encouraging the nervous impulse to travel faster.

The nervous signal actually travels only along the gaps in the myelin called **nodes of Ranvier**. Along the axon membrane at the nodes, an **action potential** occurs. An action potential involves the sequential movement of positively charged ions into and out of the neuron, which is why the signal is considered both electrical and chemical. Recall that the sodium-potassium pump is utilized by nerve cells to move three sodium ions ($Na^+$) out of the cell for every two potassium ions ($K^+$) moved in. This and other negatively charged cytoplasmic substances establish a negative **membrane potential** at rest.

As $Na^+$ rushes into the cell through ion channels, the negative membrane potential of the nerve at rest now becomes positive. This is called **depolarization**. Those channels then close, and $K^+$ rushes out of the nerve, allowing **repolarization** and the return to a negative membrane potential. A slight overshoot occurs, known as **hyperpolarization**, when the membrane potential dips even below the resting potential. The sodium-potassium pump returns the ionic balance back to homeostasis during this **refractory period**, and then the nerve is eventually ready to receive the next message.

When the action potential reaches the axon terminal, it is transferred to a truly chemical message in the form of a **neurotransmitter**. Vesicles containing the neurotransmitter are released from the neuron via exocytosis and into the **synapse**. In this space, the neurotransmitter molecules diffuse across and are picked up by receptor molecules. If these receptor molecules are part of a muscle, then a contraction is generated. If they are instead part of an endocrine gland, a different kind of chemical messenger called a hormone is released.

The glands of the endocrine system are scattered throughout the body and are integrated through the bloodstream (see Figure 18.10). The region in the inner brain known as the **hypothalamus** provides for ultimate endocrine control, detecting many major body functions. When something is out of balance in the body, the hypothalamus secretes hormones directly to the **pituitary gland** that hangs just below. Considered the master gland of the endocrine system, the

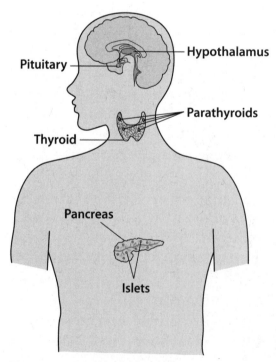

**Figure 18.10** Some Endocrine Glands

pituitary then releases one of many possible hormones into the bloodstream, where it travels to meet its target gland. When received, the target then secretes its own hormone to reestablish homeostasis.

An example of an important endocrine hormone is **insulin**, secreted by the pancreas and used to move glucose from the bloodstream into the body's cells when levels are too high. There it can be used for cellular respiration. Like most regulation, this relies on negative feedback, the ability to bring a condition back to normal when it varies. That means that another hormone must also be involved in the insulin loop to move glucose into the bloodstream from storage in the liver when levels are too low. That, in fact, is the job of another pancreatic hormone, **glucagon**.

## Female and Male Perspectives: The Reproductive Systems

Not only are the endocrine hormones involved in maintaining homeostasis in the body as described above, but they are also responsible for orchestrating sexual development and for the production and maturation of gametes. Specifically, the female hormones include **estrogen** and **progesterone,** while the main male hormone is **testosterone.** The female and male reproductive systems are specialized for the transfer of those gametes as well.

The female reproductive system is more complex than the male reproductive system in that it coordinates two separate hormone cycles and must also prepare for the potential implantation of the embryo if a fertilization event does occur. The **ovarian cycle** functions to mature a **follicle** after meiosis into an egg. **Ovulation** describes the point at which the one of the **ovaries** releases the mature egg into the **oviduct**. From there, it slowly makes its way toward the central **uterus**. If the egg is fertilized by a sperm before it reaches the uterus where it needs to implant, then it is followed by mitosis to create the very early embryo, at this point a ball of cells. Only when the developing embryo implants itself in the thickened lining of the uterus does a **pregnancy** technically occur. If the egg is not fertilized by a sperm in time, it travels out of the female reproductive tract through the **vagina** and is lost. This signals the **menstrual cycle** and the shedding of the uterine lining, no longer needed for implantation of an embryo.

The male reproductive system is focused on the production and maturation of sperm within the **seminiferous tubules** of the **testes**. The millions of sperm mature in coiled tubes just beyond called the **epididymis**. Once mature, they are moved through the muscular tube known as the **vas deferens** when signaled for release during ejaculation. Before release, the sperm are mixed with secretions from the seminal vesicles, the prostate gland, and the bulbourethral gland. Collectively these secretions nourish the sperm and provide lubrication for more effective transfer to the female via the **penis**.

**EXERCISE**
**18·1**

**Vocabulary Building.** *Explain the relationship between the following pairs of vocabulary terms.*

1. axial skeleton, appendicular skeleton

_____

_____

2. exocrine gland, evaporative cooling

_____

_____

3. pulmonary circuit, systemic circuit

_____

_____

4. nephron, ureter

_____

_____

5. axon, myelin sheath

_____

_____

**EXERCISE**
**18·2**

**Multiple Choice.** *Select the best response from the options provided to answer each question or to complete each statement.*

1. Of the following, which is the smallest level of biological organization seen specifically in the muscular system?
   a. fascicle
   b. muscle
   c. muscle tissue
   d. muscle fiber

2. Which of the components of blood functions in the blood-clotting process that follows an injury to a vessel?
   a. erythrocytes
   b. platelets
   c. leukocytes
   d. B cells

3. Which of the following would be expected to carry oxygenated blood?
   a. right ventricle
   b. pulmonary artery
   c. aorta
   d. vena cava

4. Which of the following is *not* used in chemical digestion?
   a. salivary amylase
   b. pepsin
   c. lipase
   d. progesterone

5. The subdivision of the nervous system that is directly involved in the contractions of the stomach is the
   a. somatic
   b. central
   c. autonomic
   d. sympathetic

6. Communication from a neuron to a muscle involves the action of a
   a. neurotransmitter                    c. dendrite
   b. sodium-potassium pump               d. synapse

7. A pregnancy occurs when the very early embryo implants into the wall of the
   a. oviduct                             c. uterus
   b. ovary                               d. vagina

**EXERCISE 18·3**

**Short Answer.** *Write brief responses to the following.*

1. An allergy is an undesired inflammatory response to something that is actually not harmful in any way to the body. Why might someone who suffers from an allergy require an antihistamine to combat the attack?

_____

_____

_____

_____

_____

_____

2. Explain the principles of negative feedback using the regulation of blood sugar levels by the pancreatic hormones.

_____

_____

_____

_____

_____

_____

3. Hiccups result from quick, repeated contractions of the diaphragm. Explain how this muscular contraction is related to breathing, and thus hiccups.

_____

_____

_____

_____

_____

_____

**Labeling Diagrams.** *Fill in the blanks using the following terms to correctly label the diagram representing the cavities found in the human body.*

abdominal

cranial

pelvic

thoracic

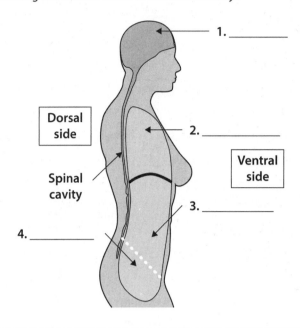

1. _____

Dorsal side

2. _____

Ventral side

Spinal cavity

3. _____

4. _____

**Thinking Thematically.** *For each of the following themes of biology, choose a different concept from this chapter and explain how it provides a useful illustration of that theme.*

1. energy and organization

_____

_____

_____

_____

_____

_____

2. form facilitates function

_____

_____

_____

_____

_____

_____

_____

3. regulation and feedback

_____

_____

_____

_____

_____

_____

## For Further Inquiry

Use the Internet to research a disorder or disease of each of the human body systems. Relate the structural problem within each system to the disruption of function.

# SYNERGY OF LIFE

# Organisms in Their Physical Environment

## Ecology

**Ecology**, the branch of biology concerned with the study of the interactions of organisms in their physical environment, has been revived lately as people's concerns with environmental conservation and the "green" movement continue to grow. Recall that organisms of the same species living together in a defined geographical region constitute a population. Populations of different species living together make up a biological **community**, and that community of organisms interacting with each other and their physical environment comprises an **ecosystem**. An ecosystem therefore includes not only the **biotic**, or living, factors like food availability and pathogen exposure but also the nonliving, **abiotic** factors like temperature and soil quality. All of the ecosystems recognized on Earth are collectively described as the **biosphere**, the living portion of the planet.

Within any ecosystem, many smaller, local habitats exist. Each **habitat** provides a specific organism with all of the resources necessary to carry out life. In the confines of that habitat, the organism is described as occupying a **niche**, an ecologically determined role that does not overlap exactly with that of another organism. This allows all organisms to coexist in an ecosystem without overstressing any one component unnecessarily. When the external environment of a habitat changes enough to put stress on an organism, it will undergo **acclimation** to physiologically adjust conditions internally within a reasonable range. For example, when the human body is exposed to high altitudes and low oxygen levels, it produces more red blood cells in an attempt to become more efficient at capturing the little oxygen available in the air. If conditions become severe, as in extreme winters or periods of heat or drought, organisms may hibernate or undergo a similar period of **dormancy**. Others may instead choose to **migrate**, or move to another more favorable region for a period of time.

## Energy Flows Through Ecosystems

Recall that energy can be transferred from one form to another and from one organism to another, but energy transfer is not a perfect process. Energy is often lost as heat, and some energy is usually invested by an organism in order to make or obtain its food. Organisms are constantly using energy to maintain life processes like metabolism because the natural tendency is toward **entropy**, or a disordered state. (Imagine a clean room quickly becoming messy with use and no effort invested into maintenance.) Energy in an ecosystem is depicted through an **energy pyramid** with various **trophic levels**, characteristic feeding roles that organisms play (see Figure 19.1).

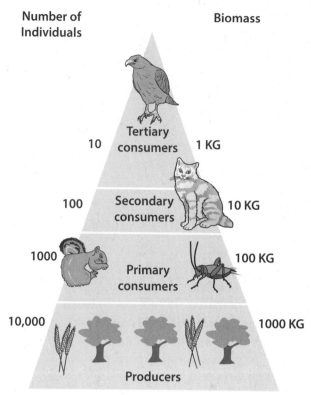

Number of Individuals

Biomass

| | | |
|---|---|---|
| 10 | Tertiary consumers | 1 KG |
| 100 | Secondary consumers | 10 KG |
| 1000 | Primary consumers | 100 KG |
| 10,000 | Producers | 1000 KG |

**Figure 19.1** Energy Pyramid

In any ecosystem, most organisms are classified as producers or consumers. The autotrophic **producers** like plants and algae are described as occupying the lowest trophic level in the energy pyramid, as they acquire energy directly from the sun via photosynthesis and do not need to feed on another organism. The producers in an ecosystem collectively possess the most **biomass** of any trophic level. The **primary productivity** of organisms at this level provides a measure of the energy available to organisms at higher levels in the pyramid.

Animals that actively eat other organisms for food are known as **consumers**. **Primary consumers** eat the producers directly and are thus **herbivores**; they are positioned at the second trophic level just above the producers. **Secondary consumers** eat the primary consumers and are thus **carnivores**; if they also directly dine upon the producers, then they would instead be considered **omnivores**. The secondary consumers are positioned at the third trophic level above the primary consumers. Tertiary and quaternary consumers may also be present in a given ecosystem and rely on the various trophic levels below them, but after that point it becomes energy-inefficient to sustain additional levels of life. That is because, according to the **10 percent rule**, a maximum of 10 percent of the energy contained within a given trophic level's biomass is actually transferred to the organisms at the next level. The rest is lost as heat, used to find and digest food, and otherwise used metabolically.

Another important group of feeders, the **detritivores**, or decomposers, comprise fungi, bacteria, and some animals like insects and worms. They are typically not depicted in an energy pyramid because they are not actively feeding but are an important element of a food chain or food web. A **food chain** is a depiction of a specific sequence of organisms involved in feeding relationships in a given ecosystem. A **food web** is then used to demonstrate the interactions and intersections of all of the food chains in that ecosystem (see Figure 19.2).

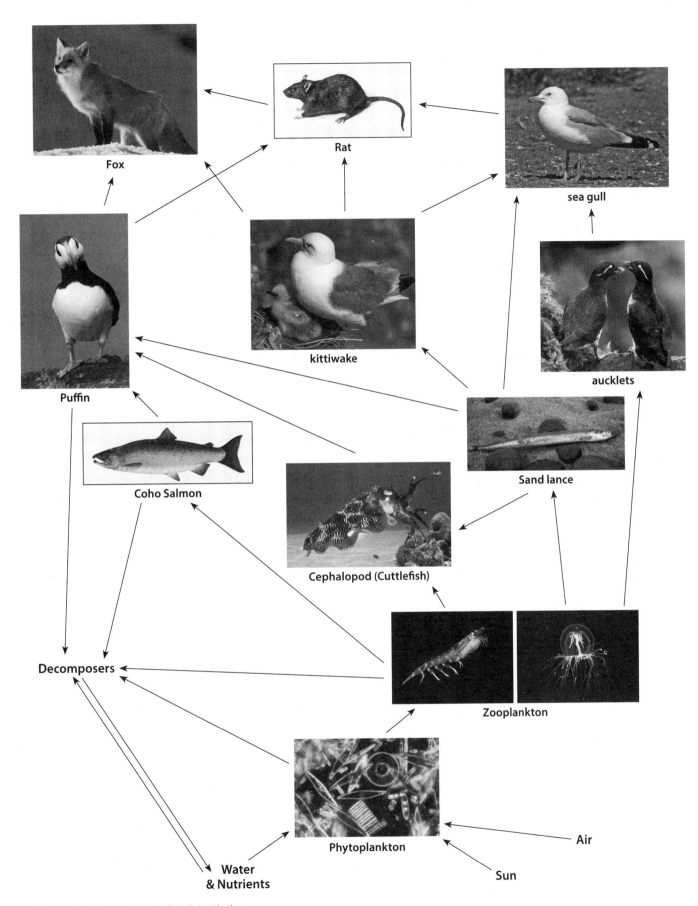

Fox

Rat

sea gull

kittiwake

aucklets

Puffin

Coho Salmon

Cephalopod (Cuttlefish)

Sand lance

Zooplankton

Decomposers

Phytoplankton

Air

Sun

Water & Nutrients

**Figure 19.2** A Marine Food Web in Alaska

# Nutrients Are Recycled Within Ecosystems

Although energy dissipates through the feeding process, nutrients are more effectively and efficiently recycled within an ecosystem. Various **biogeochemical cycles** demonstrate the mechanisms by which essential nutrients are passed between organisms and the abiotic environment. The **water cycle** is likely the most familiar to us, as we observe the various forms of water precipitation in rain, snow, and ice and also understand that organisms like plants, animals, and ourselves require a constant aqueous input (see Figure 19.3). Water not only cycles abiotically in between the atmosphere and oceans, freshwater sources, and groundwater supply by means of the dual processes of precipitation and evaporation but also is transferred biotically from plants through the transpiration process and from animals when performing evaporative cooling, breathing, and excretion. The water cycle allows organisms access not only to the essential nutrient water but more specifically to a source of hydrogen and oxygen atoms, two of the four elements most significant to life and found in every type of biological macromolecule.

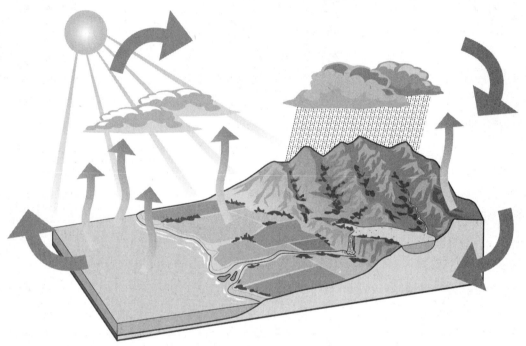

**Figure 19.3** The Water Cycle

Create a caption for Figure 19.3 to identify the specific transformation of water represented by each arrow in the water cycle.

The **carbon cycle** is best described in context of the metabolic processes of photosynthesis and cellular respiration (refer to Figure 19.4). Recall that as plant cells perform photosynthesis in their chloroplasts, they produce carbon-containing glucose for energy. When animals consume plants, the glucose is transferred to the animal. Animals break down this glucose to extract ATP and release $CO_2$ as a waste product. The carbon is thus transferred back into the atmosphere, now available for plants to absorb for the next round of photosynthesis.

The carbon cycle is, unfortunately, out of balance in the present day because of unnatural sources of $CO_2$ contributed to the atmosphere by human activity. Not only are industrial combustion and the burning of fossil fuels pumping vast quantities of $CO_2$ into the air as pollution, the massive deforestation efforts still at work around the globe result in the removal of the very autotrophs that absorb $CO_2$ and thus relieve some of the excess we have produced. Carbon, like hydrogen and oxygen from the water cycle, is necessary for every biological macromolecule, and the carbon cycle thus represents an essential ecosystem process. Efforts continue to reestablish balance within the cycle and will be discussed in more detail in Chapters 20 and 21.

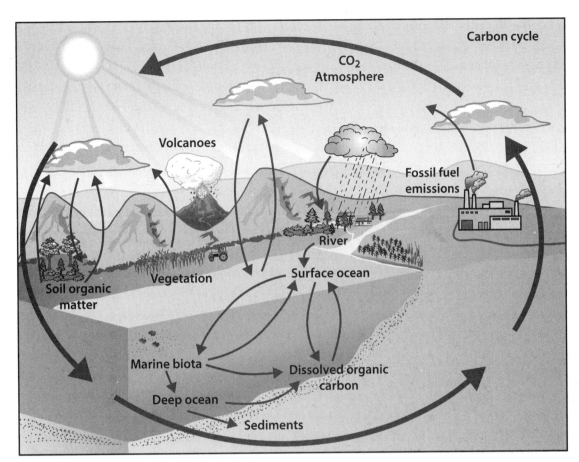

Figure 19.4 The Carbon Cycle

In addition to the water and carbon cycles, the nitrogen and phosphorous cycles complete the primary geochemical cycle picture. The **nitrogen cycle** is predominantly a soil process of **nitrogen fixation**, as various types of bacteria convert the plentiful atmospheric nitrogen ($N_2$) into useable ionic forms. These nitrogen-containing ions called nitrates ($NO_3^-$) and nitrites ($NO_2^-$) are then absorbed by plants as they take up water in their roots and passed along to other organisms through the food web. When organisms excrete, they release a nitrogenous waste in the form of ammonia ($NH_3$), uric acid, or urea, and other soil bacteria work on converting that waste again into a more useable form.

The **phosphorous cycle** provides organisms with the element that is a necessary component for ATP and nucleic acids. The element phosphorus is obtained by plants through the soil in the form of phosphate ions ($PO_4^{3-}$) and is again passed on to others through the food web. It is returned to the soil through decomposition of dead organisms and through various geological processes like weathering of rocks.

EXERCISE
19·1

**Vocabulary Building.** *Explain the relationship between the following sets of vocabulary terms.*

1. community, ecosystem

2. habitat, niche

_____

_____

3. biotic, abiotic

_____

_____

4. producer, consumer

_____

_____

5. food chain, food web

_____

_____

**EXERCISE**
**19·2**

**Multiple Choice.** *Select the best response from the options provided to answer each question or to complete each statement.*

1. Of the following biological levels of organization important to ecology, which is the least inclusive?

   a. population              c. community
   b. ecosystem               d. biosphere

2. Which of the following types of feeders is least like the others?

   a. herbivore               c. omnivore
   b. carnivore               d. detritivore

3. The ecological group of organisms primarily responsible for nitrogen fixation are

   a. land plants             c. soil bacteria
   b. soil fungi              d. algae

4. Animals contribute to the water cycle through all of the following processes *except*

   a. evaporative cooling     c. respiration
   b. transpiration           d. excretion

5. To obtain the essential elements necessary for production of ATP and nucleic acids, organisms rely on the

   a. water cycle             c. nitrogen cycle
   b. carbon cycle            d. phosphorous cycle

**Short Answer.** *Write brief responses to the following.*

1. Explain the following statement: *energy flows through an ecosystem, while nutrients are cycled within it.*

   _____

   _____

   _____

   _____

   _____

   _____

2. In an ecosystem, the secondary consumers possess 1,000,000 units of energy within their biomass. Use the 10 percent rule to determine the maximum quantity of energy that the tertiary consumers would themselves possess.

   _____

   _____

   _____

   _____

   _____

   _____

3. What do you observe in your everyday life that contributes to the imbalance of the carbon cycle? What can you personally do to help reestablish equilibrium?

   _____

   _____

   _____

   _____

   _____

   _____

**Labeling Diagrams.** *Examine the following diagram representing a complex marine food web. Use the information in the diagram to answer the questions that follow.*

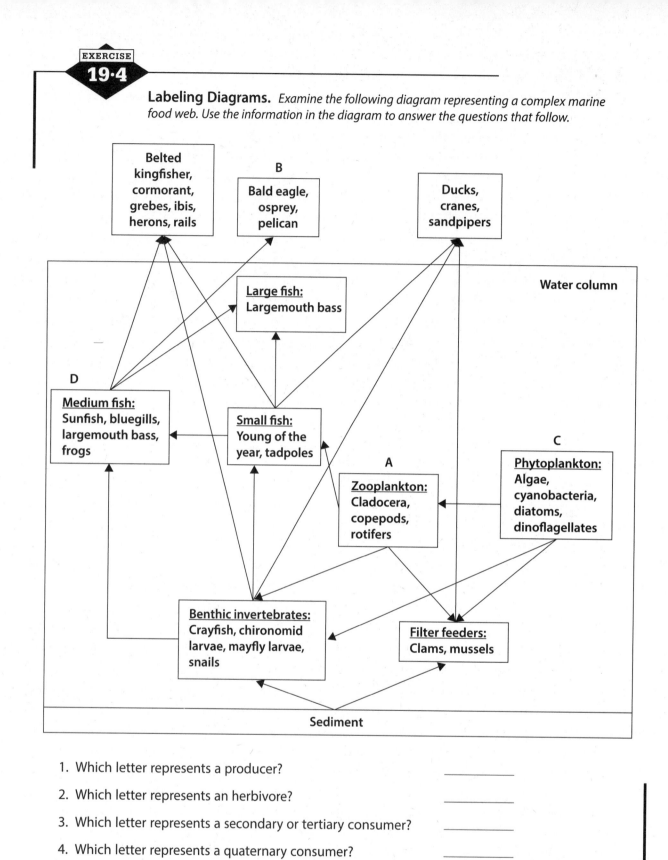

1. Which letter represents a producer? _____

2. Which letter represents an herbivore? _____

3. Which letter represents a secondary or tertiary consumer? _____

4. Which letter represents a quaternary consumer? _____

**EXERCISE 19·5**

**Thinking Thematically.** *For each of the following themes of biology, choose a different concept from this chapter and explain how it provides a useful illustration of that theme.*

1. energy and organization

_____

_____

_____

_____

_____

_____

2. natural interdependence

_____

_____

_____

_____

_____

_____

3. regulation and feedback

_____

_____

_____

_____

_____

_____

## For Further Investigation

Conduct some online research on deep-sea vent ecology. Create a sample food chain that includes the following: chemosynthetic bacteria, tube worms, black smoker, and spider crabs. Now consider the natural ecosystems where you live. Find a substitute organism that serves in the same trophic level as each named in the deep sea vent ecosystem.

# Organisms Living Together

## Populations and Communities

No organism lives in isolation. Even solitary animals exist together in a larger region with other members of the same species, constituting a population. Populations of different species overlap and interact in natural habitats, forming a biological community. These dynamic populations and communities are constantly responding to a changing environment, as daily and seasonal cycles affect the way that organisms access the abiotic and biotic resources necessary for the continuation of life.

## Population Factors: Indicators of Health

In a rainforest in Brazil, a population of frogs begins to decline sharply in number, and local ecologists become concerned. What factor or factors may be responsible here? Ecologists may investigate data regarding **population density**, the number of individual organisms per unit area (in terrestrial habitats) or per unit volume (in aquatic ones). Not only is the density significant, for over-dense populations may suffer from resource shortage and increased risk of disease, but the **dispersion** of the organisms throughout their habitat is also significant. If organisms exhibit **uniform distribution**, their even allotment likely places less stress on their surroundings compared to organisms that exhibit **clumped distribution** and are very highly concentrated in relatively few areas. Ecologists will also need to know the average **birth rate**, the number of new organisms that join the population per unit of time, and the **death rate**, the number of deaths experienced by members of the population over time. The typical **life span** of an organism, or the average life expectancy, may also be helpful.

If the frog population had initially been experiencing many more births than deaths, then the **growth rate** may have been unsustainable. The population may have reached its **carrying capacity**, the maximum number of individuals that can be supported adequately by the local ecosystem. Many populations of organisms demonstrate **exponential growth** when resources are plentiful and organisms are well dispersed throughout the ecosystem (see Figure 20.1). Eventually, however, certain limiting factors begin to present problems for some of the members of the population. Predation, disease, and availability of and competition for resources are just some potential factors that affect the survival and reproductive capabilities of a population. Factors whose effects vary with the density of organisms present are called **density-dependent factors**. Examples include mainly biotic factors such as competition for mates, overcrowding, and disease. If instead the factor affects the population regardless of its density, then it is called a

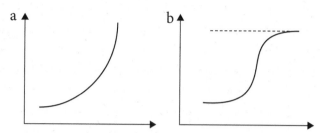

**Figure 20.1** Exponential and Logistic Growth Curves

**density-independent factor**. Density-independent factors include predominantly abiotic factors like sunlight availability, ambient temperature, and the unintended consequences of human activity.

Most populations in realistic ecosystems tend to demonstrate **logistic growth** (see Figure 20.1) instead of exponential growth. While the initial shape of the curve is similar, the pressure experienced from limiting factors begins to slow the rate of increase. The overall logistic curve is described as S-shaped, with the upper plateau representing the carrying capacity. The exponential growth curve is instead J-shaped.

# Community Factors: Indicators of Health

Natural ecosystems of course do not encompass only one species in isolation but rather are true biological communities. Just as key factors are measured to assess the health of populations, factors like species richness and species evenness are used for evaluating communities. **Species richness** refers to the number of different species present in a given community, while **species evenness** assesses the relative abundance of different members of the community. Together, these measures provide an approximation of the **biodiversity** of the region, the overall variation of life present, and an important indicator of the health of the community.

In a community, members of different species are often interacting with one another in one of three types of **symbiosis**. **Mutualism** indicates an interaction that is positive to both members involved (+/+). Often organisms involved in mutually beneficial relationships undergo coevolution because of their extreme interdependence. **Commensalism** is a symbiotic relationship in which one organism benefits, while the other organism is neutrally affected (+/0). Finally, **parasitism** describes a symbiosis in which one organism benefits while the other is harmed in some way (+/–).

Communities that have experienced some form of **disturbance**, or change that affects the survival of the various species present, typically undergo a predictable sequence of events called **ecological succession** to reestablish a stable, biodiverse community (see Figure 20.2). Succession is characterized first by the presence of **pioneer species**, organisms that are hardy, fast-growing, and disperse seeds easily. If a community is being established for the first time in a given area where no soil is present, then **primary succession** is occurring. This can be observed in nature when volcanic eruptions deposit a fresh lava bed or when a glacier retreats and exposes new land. It tends to be slow relative to **secondary succession**, which occurs when the disturbance is instead experienced in a region where existing soil is still present. This can be seen when a hurricane devastates a coastal region or when humans deforest a hardwood forest for industry. When ecological succession ends, the resulting organisms constitute a **climax community**, one that is mature, stable, and healthy. Succession never really ends, however, for a disturbance of some sort ultimately occurs, and the process starts anew.

**Figure 20.2** Ecological Succession in a Florida Ecosystem

**Vocabulary Building.** *Provide a definition for each of the following vocabulary terms. When possible, identify any roots in the term and use them to help create the definition.*

1. population density

_____

_____

2. carrying capacity

_____

_____

3. logistic growth

_____

_____

4. symbiosis

_____

_____

5. ecological succession

_____

_____

**Multiple Choice.** *Select the best response from the options provided to answer each question or to complete each statement.*

1. Which of the following is an example of a density-independent factor relevant to a population of giant kelp in the Pacific Ocean off the coast of California?

    a. water temperature

    b. precipitation levels

    c. herbivore population levels

    d. presence of a parasitic fungus

2. A bird inhabits a natural hole in the trunk of a tree; the bird consumes a worm that likes to eat the new leaves of the tree. This is an example of

    a. commensalism

    b. parasitism

    c. vitalism

    d. mutualism

3. A population of bacteria decomposing a dead horse is likely to demonstrate

    a. a J-shaped growth curve

    b. an S-shaped growth curve

    c. exponential growth

    d. both a and c

4. Which of the following is *least* likely an example of interspecific competition?

    a. competition between predator and prey

    b. competition for mates

    c. competition for water sources

    d. competition for safe habitat

5. The lava cactus is a pioneer species found in the volcanically active Galapagos Islands. One would expect the plant to have all of the following characteristics *except*

    a. fast growth

    b. well-dispersed seeds

    c. slow reproduction rates

    d. simple nutritional needs

**Short Answer.** *Write brief responses to the following.*

1. Which is more significant in assessing the health of a given population, population density or dispersion? Explain.

_____

_____

_____

_____

_____

_____

2. How do species richness and species evenness provide a measure for the biodiversity of a community?

_____

_____

_____

_____

_____

_____

3. Provide an example of a disturbance that might lead to primary succession and one that would cause secondary succession. How are both types related to a climax community?

_____

_____

_____

_____

_____

_____

_____

**Interpreting Diagrams.** *Analyze the following diagram representing hypothetical growth curves. Use the information in the diagram to answer the questions that follow.*

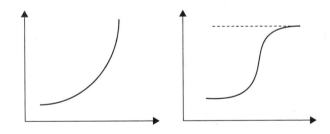

1. Which curve is likely observed for a colony of bacteria growing in a Petri dish?

_____

2. The dashed line in the second curve is representing what?

_____

**Thinking Thematically.** *For each of the following themes of biology, choose a different concept from this chapter and explain how it provides a useful illustration of that theme.*

1. regulation and feedback

_____

_____

_____

_____

_____

_____

2. continuity and change

_____

_____

_____

_____

_____

_____

3. science methodologies and applications to society

_____

_____

_____

_____

_____

_____

## For Further Investigation

Use the Internet to investigate an invasive species, one that has been introduced to a community (knowingly or unknowingly) by humans and has no natural predators. What type of growth does that population demonstrate? What is the effect of the invasive species on the natural populations of organisms living there?

# Earth and the Human Factor

## The Biosphere

The first German astronaut, Sigmund Jähn, captured so eloquently the challenge of the environmentalist movement when he said, "Before I flew I was already aware of how small and vulnerable our planet is; but only when I saw it from space, in all its ineffable beauty and fragility, did I realize that humankind's most urgent task is to cherish and preserve it for future generations." While we all can't have a firsthand account of the view of Earth from space, we all should imagine and retain that perspective. Only then can we really appreciate the enormity of the **biosphere**, the living portion of the planet Earth. The biosphere extends from the depths of the oceans to the peaks of the highest mountains, reaching well into regions that may at first seem inhospitable to life. The biosphere is organized into **biomes**, expansive areas of land or volumes of water with characteristic climates and displays of biodiversity (see Figure 21.1).

## Terrestrial Biomes

In the northernmost regions of the globe, the terrain is blanketed in snow or ice for the majority of the year. In fact, the forever-frozen soil called **permafrost** is characteristic of this **tundra** biome. Only organisms well adapted to such environmental extremes can tolerate the tundra, and many are only part-time migratory residents of the region. The **coniferous forest**, also known as the **taiga**, dominates regions just to the south and is characterized by large, needle-bearing conifers as the major ecosystem producers.

Continuing farther south, temperate forests and grasslands can be found. **Temperate forests** are composed of mainly deciduous trees that lose their leaves in the winter when the region becomes seasonably cold, while **grasslands** possess only smaller, scattered trees and shrubs. In areas with significantly less annual rainfall, chaparral and deserts can be found. **Deserts** demonstrate more climatic extremes than **chaparral**, which is characteristic of the more favorable weather typical of the Mediterranean and of Southern California.

In equatorial regions where sunlight remains relatively constant throughout the year, the **tropical forests** take hold. Characterized by excessive rainfall, high nutrient availability, and warm temperatures, the tropical forests support the highest concentration of biodiversity on terrestrial Earth. Specialized plants called epiphytes adapted to survive in the highly competitive environment where the thick **canopy** of treetops blocks most of the sun from ever reaching the forest floor. These **epiphytes** possess the unique ability to "take root" on top of the branches and trunks of other trees, never actually making contact with the soil.

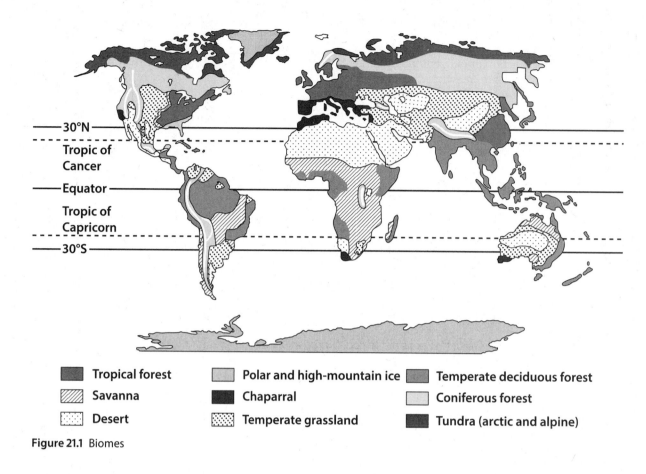

Tropical forest   ▢ Polar and high-mountain ice   ▢ Temperate deciduous forest
Savanna          ■ Chaparral                       ▢ Coniferous forest
Desert           ▒ Temperate grassland             ▢ Tundra (arctic and alpine)

**Figure 21.1** Biomes

## Aquatic Ecosystems

Recall that at least 70 percent of the surface of the Earth is covered in water, so aquatic biomes are at least as significant as the terrestrial ones we tend to be more familiar with. Access to sunlight becomes a much more significant concern to organisms in these ecosystems because certain wavelengths of light can only penetrate a relatively shallow region under water. The region that does receive sunlight for photosynthesis is known as the **photic zone**, while the dependent region below without access to light is instead called the **aphotic zone** (see Figure 21.2). Unicellular algae constitute the majority of producers within the oceanic plankton and thus support the rest of the organisms in the ecosystem. While mostly applicable to marine ecosystems in oceans and seas, the terms *photic* and *aphotic* are

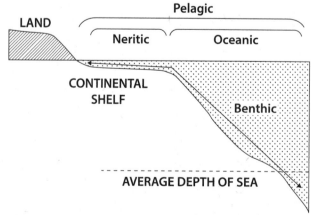

**Figure 21.2** Ocean Zones

also relevant to organisms living in really large, deep freshwater lakes. Regardless of depth or location, the bottom surface and sediment of any aquatic ecosystem is referred to as the **benthic zone**.

Where ocean meets land, the **intertidal zone** exists. Organisms that occupy this region must be able to tolerate daily extremes, as they are cyclically exposed to turbulent water and then dry air, like those inhabiting tide pools. Beyond the intertidal zone, the bulk of the ocean waters constitute the **pelagic zone**. The portion of the pelagic zone closest to land and sitting above the continental shelf is referred to as the **neritic zone**, while the true open ocean portion characterizes the **oceanic zone**. Only in the extreme depths of the ocean does the **abyssal zone** exist, a unique region characterized by deep-sea vents and chemosynthetic bacteria.

Specialized marine ecosystems include the coral reefs and estuaries. **Coral reefs** represent the most biodiverse aquatic ecosystems, comprising countless varieties of **zooxanthellae**, partnerships between corals and their photosynthetic algal endosymbionts. **Estuaries** are areas where **rivers** and streams carrying freshwater runoff from high elevations meet the ocean, so they present a unique mix of nutrients and other environmental conditions.

Freshwater ecosystems are less prevalent than their marine counterparts but are still of great significance in supporting life. **Wetlands** are at the boundary of terrestrial and aquatic biomes, for they consist of land with significant portions saturated with pockets of freshwater. Freshwater lakes can vary greatly depending on size and composition. **Eutrophic lakes** are characterized by murky waters rich in nutrients and more supportive of life, while **oligotrophic lakes** are composed of clear waters lacking in essential resources.

# Human Impact and the Green Movement

In the middle of the 20th century, conservationist and marine biologist Rachel Carson attempted to warn the world about the potentially catastrophic future we would face should we not collectively change behaviors that were negatively affecting the planet. Most notably, her book Silent Spring brought major attention to the ways in which pesticides like DDT were harming populations of organisms. Carson not only helped to curb pesticide use throughout much of the world, she is credited with helping launch the modern environmental movement.

The pollution caused by human activity and intensive industry, and exacerbated by the unmatched population explosion our species has experienced, has contaminated the essential resources within all nonliving portions of the Earth. The **lithosphere**, the upper portions of the Earth's crust, is contaminated with pesticides, fertilizers, and other potential biotoxins. The **hydrosphere**, the portion of the planet covered in water, includes areas of extreme pollution like the several enormous plastic-garbage patches that have been created in the oceans where currents converge. Various freshwater sources have been contaminated by **acid precipitation**, and drinking water supplies are threatened by seepage from contaminated soil. Finally, the **atmosphere** has experienced significant and worrisome changes, as the industrial and residential emissions of **greenhouse gases** like carbon dioxide, methane, and nitrous oxide act to trap solar radiation close to the Earth's surface. Evidence for contamination of the atmosphere ranges from global warming and climate change to depletion of chunks of the protective **ozone layer** to the industrial **smog** that blankets many overpopulated, industrial cities.

Although humans are undoubtedly the species responsible for the disproportionate number of local and total extinction events that currently occur, for the massive loss of biodiversity experienced on Earth, and for the disruption and destruction of countless habitats of many other species, we also have the most power to revive and restore the planet. **Conservation biology** and **restoration biology** are concentrated on studying ways to preserve natural lands, resources, and biodiversity—and also how to replenish such biodiversity when lost. The **environmental sciences** take a slightly larger scope and focus increasingly on the abiotic contributions to the biosphere. But more importantly, all citizens of the globe are increasingly being encouraged to get involved. **Ecotourism** seeks to expose travelers to new lands while promoting and preserving the local ecosystem and supporting the local economy.

More recently, city dwellers are encouraged to take part in **urban ecology**. Central Park in New York City, for example, represents a great place for such exploration of regions where people and natural ecosystems interact in novel ways. The everyday observer of nature is encouraged to take part in ecological surveys, often employing handheld technologies to record data on various species in the biological community or to report potential environmental concerns and thus prompt more immediate action. Education and advocacy efforts attempt to arm the urban ecologist with a practical toolkit of knowledge and skills to become an active participant in the preservation of their local communities.

Arguably, the most important thing that individuals can do to help restore homeostasis within the biosphere is to find ways to live **sustainably**, without placing disproportionate strain on natural ecosystems and resources, and to encourage similar actions in others. This is a goal of the recent green movement that has captured the attention of many globally. Renowned American biologist E. O. Wilson has estimated that if all humans on Earth consumed at the rate Americans do, we would require at least four more Earth-equivalents to provide the resources to support that life. As disheartening as that at first sounds, it also presents a very powerful motivation for action and change moving forward.

EXERCISE
**21·1**

**Vocabulary Building.** *Explain the relationship between the following pairs of vocabulary terms.*

1. tundra, permafrost

_____

_____

2. epiphyte, canopy

_____

_____

3. photic zone, plankton

_____

_____

4. wetlands, estuary

_____

_____

5. greenhouse effect, global warming

_____

_____

**Multiple Choice.** *Select the best response from the options provided to answer each question or to complete each statement.*

1. The terrestrial biome characterized by marked climatic seasonal changes and large stands of trees with leaves that periodically change color and fall off is the
   a. temperate forest
   b. coniferous forest
   c. tropical forest
   d. tundra

2. In aquatic ecosystems, which zone does not receive light?
   a. photic
   b. benthic
   c. aphotic
   d. both b and c

3. Which of the following represents a unique symbiotic relationship seen in coral reef ecosystems?
   a. epiphytes
   b. lichens
   c. zooxanthellae
   d. permafrost

4. A lake that is teeming with life and resources would be characterized as
   a. photic
   b. eutrophic
   c. aphotic
   d. oligotrophic

5. An individual can help reduce the strain placed on the biosphere by taking which of the following actions?
   a. studying environmental sciences
   b. participating in urban ecology
   c. using resources sustainably
   d. all of the above

EXERCISE
21·3

**Short Answer.** *Write brief responses to the following.*

1. Compare and contrast the pelagic zone and the intertidal zone.

_____

_____

_____

_____

_____

_____

2. In addition to producing more greenhouse gases, humans are also participating in deforestation activities around the globe. Some practices are more sustainable than others, but in any regard, deforestation only exacerbates the global warming problem. Considering the ecological significance of plants, explain how this is the case.

_____

_____

_____

_____

_____

_____

**Interpreting Diagrams.** *Examine the following diagram representing a food chain and the potential effects of biological magnification of a toxin. Use the information in the diagram to answer the questions that follow.*

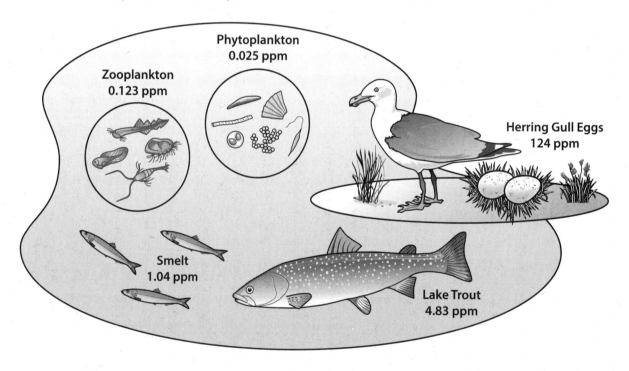

1. How many times greater is the concentration of toxin in the trout versus the smelt?

_____

2. If the toxin becomes deadly at 100 ppm (parts per million) for a mammal, how many trout would the mammal have to eat to suffer the effects? How many gull eggs would need to be consumed?

_____

**EXERCISE**
**21·5**

**Thinking Thematically.** *For each of the following themes of biology, choose a different concept from this chapter and explain how it provides a useful illustration of that theme.*

1. energy and organization

_____

_____

_____

_____

_____

_____

2. science methodologies and applications to society

_____

_____

_____

_____

_____

_____

3. continuity and change

_____

_____

_____

_____

_____

_____

## For Further Investigation

Devise a plan to lessen your own impact on the Earth and its precious resources. Could you eat little to no meat and eat more sustainably within the food chain? Or use resources less or more efficiently? Could you support local farms and businesses and reduce the ecological and other costs of transportation or avoid keeping and using toxic chemicals in your home or place of work? Even if you are already living "green," challenge yourself to find a new way to contribute to a sustainable Earth for all.

# Putting It All Together

At this point, you have the knowledge and tools to see connections between things in the natural world that were previously invisible. Just as the microscope advanced our understanding of bacteria and viruses by allowing the structure and behavior of these tiny infectious bodies to be observed, the intensive exploration of the living world you have now considered provides you with new and exciting vantage points for understanding yourself and all of the other organisms inside you and all around you—whether you can actually see them or not!

Using the themes of biology, you have already been asked to take a broader perspective and consider intersections between concepts that might seem at first unrelated. Now is an opportune time to apply the knowledge you have amassed in a different way: working from a broad, macroscopic level and tracking the levels of biological organization all the way back down to the microscopic and even molecular level.

Armed with this comprehensive (new or refreshed) biology knowledge, you should feel better equipped to discuss current events involving the living world, and also to help clarify common misconceptions that persist and to debunk myths that attempt to challenge the field itself.

## Global Warming

Ecological issues make for ideal examples for this type of deductive exercise. As referenced in Chapter 21, **global warming** is a critical issue that plagues the world today (see Figure 22.1).

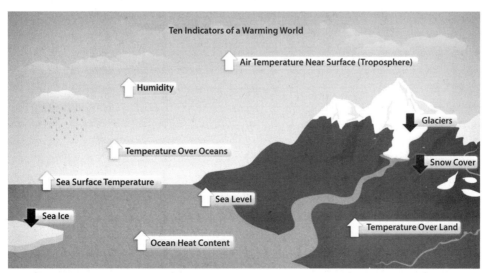

**Figure 22.1** Ten Indicators of a Warming World. (Note: Arrows indicate whether the factor is increasing [up] or decreasing [down] as climate change occurs.)

An overwhelming amount of evidence demonstrates that the temperatures of Earth's land and water surfaces, as well as the temperature of the surrounding atmosphere, have all increased markedly throughout the course of humanity. In more recent times, technologies improved at an increasingly rapid pace, accelerating the growth of human populations to exponential levels and contributing to huge increases in levels of atmospheric $CO_2$ (refer to Figure 22.2).

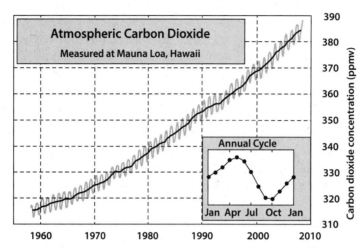

**Figure 22.2** Atmospheric Carbon Dioxide Concentrations from 1958–2008

Just like the glass walls and ceiling in a greenhouse act to trap the radiation from the sun and increase the air temperature inside, **greenhouse gases** in Earth's atmosphere warm the contents within. Although a natural phenomenon, the **greenhouse effect** has been accelerated over the past several decades. Gases like $CO_2$, methane, and water vapor are the most significant contributors (see Figure 22.3); working to decrease levels of such greenhouse gases is a key area in many current climate change initiatives.

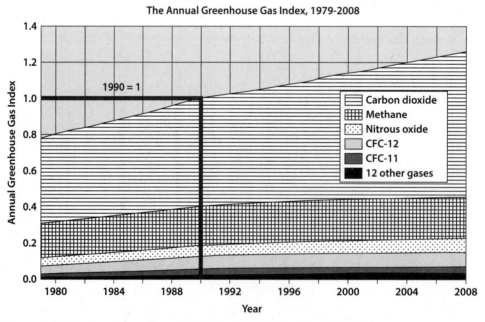

**Figure 22.3** The Annual Greenhouse Gas Index, by Gas Type from 1979–2008

Unfortunately, combustion of fossil fuels and deforestation together are responsible for the imbalances in the carbon cycle. As more and more fossil fuels continue to be combusted for energy, more and more carbon dioxide will build up in the atmosphere. At the same time, deforestation continues to supply humans with countless products generated from trees. Unfortunately, these forests act as a massive $CO_2$ sponge, absorbing this gas for photosynthesis. Without them, the $CO_2$ accumulates and acts as a greenhouse gas instead.

Global warming can affect every level of biological organization beneath it. For example, global warming affects both the population and community ecology levels by increasing the range of mosquitos and thereby making human-mosquito parasitism increasingly common. In turn, more malaria infections are likely, which plays out inside the infected individuals' circulatory systems as their bloodstream and other internal organs host the stages of the Plasmodium life cycle. During this infection, individuals' red blood cells burst in a cytolysis-like fashion, exploding with new cells.

# The Sixth Mass Extinction?

Another useful deductive example is the **biodiversity crisis** that ecosystems face worldwide. Many biologists agree that we are currently in the midst of the sixth major mass extinction ever experienced on Earth. During a **mass extinction event**, population crashes take place as extinction rates far outpace background levels for an extended period of time (refer to Figure 22.4). All of this results in the loss of at least 75 percent of Earth's biodiversity. What makes this sixth occurrence unique is that one species is primarily responsible for the extinction of so many others: our own species, *Homo sapiens*.

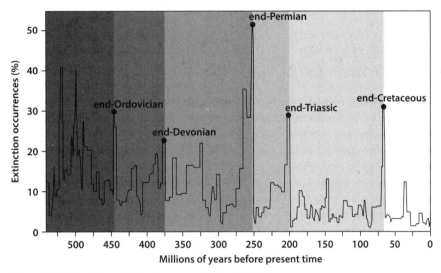

**Figure 22.4** Five Mass Extinction Events in Earth's History

We face a current biodiversity crisis due in large part to the high levels of **ecosystem degradation** that have taken place on Earth. Habitat loss for critical species in many ecosystems is caused by human activities like overpopulation and unsustainable use of resources, deforestation, disruption of waterways, and pollution of a variety of natural resources.

The presence of **invasive species** can also have a devastating effect on a local population and contribute to a population crash. If a species transported to a nonnative ecosystem by humans, is evolutionarily well fit for its new home, and has few or no local predators to keep its population regulated, it can grow unchecked, wipe out prey species, and outcompete the previous occupant of the same niche.

When any population reaches a critically small level, it risks an **extinction vortex** (shown in Figure 22.5). This takes place when mutually reinforcing abiotic and biotic processes play out in such a way as to further decrease the population size as it tends toward local extinction.

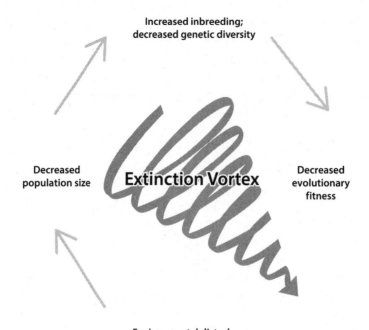

**Figure 22.5** Extinction Vortex

The smaller a population gets, the more likely it is that inbreeding and random genetic drift will lead to a reduction in genetic diversity within the population. Population and individual fitness are then affected negatively such that the population experiences lower reproductive rates and higher death rates. An even smaller subsequent generation results.

# Challenges to the Discipline

Unfortunately, **science denial** has become more prevalent in many societies (see Figure 22.6). Most of this is directed at the biological sciences, and in particular toward its core concept of evolution. This particular form of science denial is not new; when Darwin published *On the Origin of Species*, he intentionally omitted the topic of human evolution, knowing just how controversial the topic would be. Now to many the case of evolution conflicted with their religious interpretation of the origin of life and challenged their understanding of the world in fundamental ways.

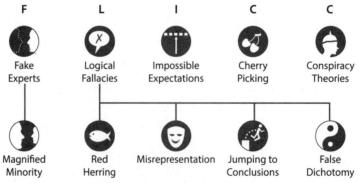

**Figure 22.6** Five Common Means of Science Denial

Interestingly there are many examples that directly demonstrate evolution in action. Plastic-digesting bacteria exist; however, their food source is synthetic and has only been around for a little over a century. Evolution explains how these new species of microbe came to be through mutation of crucial enzymatic genes. Likewise, there is concern that we will run out of crucial medications to prevent and treat certain infectious diseases because of antibiotic and antiviral resistance that is taking place. The development of genetic resistance in a population is also evolution in action!

More recently, science denial has affected other applications of biology like global warming and childhood vaccinations. While the consensus among scientific researchers is that humans are primarily responsible for the current levels of climate change we are experiencing (refer to Figure 22.7), many laypeople still readily challenge the notion.

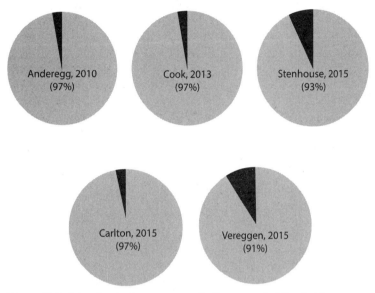

**Figure 22.7** Scientists in Consensus on Anthropogenic Climate Change
(Note: The lead study author and publication date are indicated per study.)

Additionally, a plethora of childhood vaccinations have been demonstrated time and time again to be safe and effective, not only at protecting the recipient of the vaccine, but also by passively protecting others who are too young or immunocompromised and who cannot receive it themselves. In fact, **herd immunity** has been shown to be quite effective as a vaccination technique (see Figure 22.8). With this approach, not every individual needs to be vaccinated within a population to achieve success, but a critical mass does.

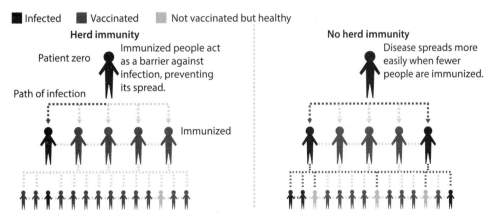

**Figure 22.8** Effects of Vaccination Through Herd Immunity

Some continue to challenge the safety and efficacy of these vaccines, but there is no scientific evidence to support such a stance. When populations become lax with their vaccination rates, huge increases in the incidence of diseases often occur (as demonstrated in Figure 22.9), and in some cases, mortality rates can increase as well.

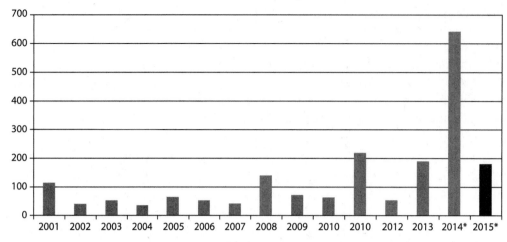

U.S. Measles Cases by Year

*Provisional data reported to CDC's National Center of immunization and Respiratory Diseases

**Figure 22.9** Annual Measles Infections in the United States from 2001-2015

Vaccines have in fact been one of the most dependable and significant contributors to increase in life expectancy in humans over the past century, along with major improvements to sanitation and hygiene. The global response that resulted in several COVID-19 vaccines being bioengineered in a year's time clearly demonstrates the utility of vaccines, especially when an easily transmitted, newly emerged virus has evolved and presents an immediate threat to human populations. Whatever the challenge or supposed controversy in science, it is helpful to consider the wisdom of world-renowned astrophysicist Neil deGrasse Tyson, who famously said, "The good thing about science is that it's true whether or not you believe in it."

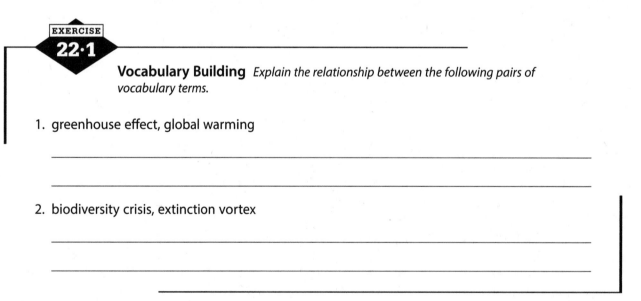

EXERCISE
22·1

**Vocabulary Building** *Explain the relationship between the following pairs of vocabulary terms.*

1. greenhouse effect, global warming

_____

_____

2. biodiversity crisis, extinction vortex

_____

_____

**Multiple Choice** *Select the best response from the options provided to answer each question or to complete each statement.*

1. Which of the following would be *least* likely to contribute to habitat loss for a particular species?
   a. urban development
   b. underuse of natural resources
   c. deforestation
   d. damming of a river

2. Invasive species
   a. usually have positive effects on local ecosystems.
   b. involve organisms that have migrated away from but return to their prior habitat.
   c. often have no natural predators, and their populations grow out of balance.
   d. are always introduced intentionally by humans.

3. Science denial can result from
   a. misrepresentation of facts
   b. conspiracy theories
   c. cherry-picking of specific ideas
   d. all of the above

**Short Answer** *Write brief responses to the following.*

1. Although devastating, mass extinctions can present new opportunities that didn't exist before in a previously crowded ecosystem. Describe two or three (hypothetical) positive outcomes of mass extinctions for those species that remain.

_____

_____

_____

_____

_____

2. Consider the issue of antibiotic resistance. How does this play out on the cellular level? On the population and community levels?

_____

_____

_____

_____

_____
_____

**Interpreting Diagrams** *Examine the following common representation of human evolution. Explain how this type of sequential representation of related primate species can be misleading and unintentionally equip challengers to evolutionary theory with misinformed counterpoints. (Hint: Consider how speciation occurs.)*

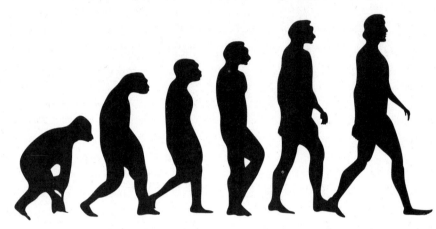

**Thinking Thematically** *For each of the following themes of biology, choose a different concept from this chapter and explain how it provides a useful illustration of that theme.*

1. continuity and change

_____
_____
_____
_____
_____
_____

2. regulation and feedback

_____
_____
_____
_____
_____
_____

3. scientific methodologies and applications to society

_____

_____

_____

_____

_____

_____

## For Further Investigation

Choose one specific area within biology that is often challenged or targeted as controversial. Outline both sides of a debate, providing specific evidence for each point as much as possible.

# Appendix

## Identifying Roots for Vocabulary Building

Use the following list of mainly Greek and Latin roots, their meanings, and the biologically relevant examples provided to help build your vocabulary. Each time a new vocabulary term is encountered throughout the text (important vocabulary terms appear in boldface), you are encouraged to identify any roots present in the term and then to assemble the meanings of those roots into your own etymological definition. You are also encouraged to keep a list of additional examples of words (biological or otherwise) that share those roots by adding your own to the list below.

| ROOT | MEANING | EXAMPLE | MORE EXAMPLES |
|---|---|---|---|
| a(n)- | not, without | anaerobic | |
| ab- | away from | abductor | |
| ad- | next to, at, toward | adductor | |
| adipos- | fat | adipose | |
| aer- | air | aerobic respiration | |
| agri- | field | agriculture | |
| alge- | sensitivity to pain | analgesic | |
| alb- | white | albinism | |
| allo- | other | allopatric speciation | |
| amnio- | fetal membrane | amniocentesis | |
| amphi- | double, both | amphibian | |
| amyl- | starch | amylase | |
| ana- | upon, upward | anaphase | |
| andros- | male | androsterone | |
| angios- | vessel, covering | angiosperm | |
| anima- | spirit, living being | Animalia | |
| anter- | front | anterior | |
| antho- | flower | antheridium | |
| anthro- | human | anthropoid | |
| anti- | opposite, against | antibody | |
| append- | to hang upon | appendicular skeleton | |
| aqua- | water | aqueous | |
| arach- | spider | arachnid | |
| arbor- | tree | arboreal | |
| arch- | to begin | archegonium | |
| archae- | ancient | Archaebacteria | |
| arthro- | joint | arthropod | |
| atri- | main room | atrium | |
| austral- | southern | *Australopithecus* | |
| auto- | self | autosome | |
| auxi- | to increase | auxin | |
| avi- | bird | Aves | |
| axi- | axle | axial symmetry | |

| ROOT | MEANING | EXAMPLE | MORE EXAMPLES |
|---|---|---|---|
| basi- | footing, base | Basidiomycota | |
| bi- | two | bilateral symmetry | |
| bio- | life | biology | |
| brachio- | arm | brachiopod | |
| bronch- | throat, windpipe | bronchus | |
| bulbo- | bulbous | bulbourethral gland | |
| caecu- | blind | cecum (caecum) | |
| calc- | limestone | calcium | |
| calor- | heat | calorie | |
| capill- | hair | capillary | |
| carcino- | cancer | carcinogen | |
| cardia- | heart | cardiac muscle | |
| catabol- | throwing down | catabolism | |
| catalys- | to loosen, to untie | catalysis | |
| cav- | hollow | cavity | |
| cephalo- | head | cephalopod | |
| cerebr- | of the brain | cerebrum | |
| chiasm- | crossing | chiasma | |
| chloro- | green | chloroplast | |
| chondr- | cartilage | Chondrichthyes | |
| chorion | outer membrane | chorionic villi sampling | |
| chord- | string, rope | chordate | |
| chrom- | color | chromosome | |
| chryso- | golden | Chrysophyta | |
| chym- | bodily juice | chyme | |
| cili- | eyelash | cilia | |
| circu- | around | circuit | |
| clad- | branch | cladogram | |
| coccus | berry, grain | *Streptococcus* | |
| coel- | hollow cavity | coelomate | |
| colon- | large intestine | colonic | |
| co-/com- | with, together | community | |
| con- | with | convergent evolution | |
| contra- | against, opposite | contraception | |
| corpus | body | corpus luteum | |
| cortex | outer rind, bark | cerebral cortex | |
| costa- | rib | costal bone | |
| cotyl- | cup-shaped | cotyledon | |
| crani- | skull | cranium | |
| -crine | to distinguish | endocrine | |
| crypt- | hidden | cryptic coloration | |
| cuti- | skin | cuticle | |
| cyano- | blue | cyanobacteria | |
| cycl- | circular | cyclin | |
| cyto- | container, basket | cytoplasm | |
| decid- | falling off | deciduous forest | |
| dem- | population | demography | |
| dendr- | tree | dendrite | |
| derm- | skin | dermis | |
| di(a)- | two | divergent evolution | |
| dia- | through, across | diagram | |
| didym- | twined | epididymis | |
| diffus- | to scatter, pour out | diffusion | |
| dinos- | terrible | dinosaur | |
| diplo- | doubled | diploid | |
| dors- | of the back | dorsal | |
| duoden- | twelve | duodenum | |
| dys- | bad, difficult | muscular dystrophy | |

| ROOT | MEANING | EXAMPLE | MORE EXAMPLES |
|------|---------|---------|---------------|
| e- | without, removed | enucleated | |
| echino- | spiny | echinoderm | |
| eco- | habitation, house | ecology | |
| ecto- | outside, external | ectotherm | |
| ejacul- | to throw out | ejaculation | |
| endo- | inside | endosymbiosis | |
| enzym- | to leaven | enzyme | |
| epi- | on, upon | epidermis | |
| erect- | upright | *Homo erectus* | |
| ergon- | work | endergonic | |
| erythr- | to be red | erythrocyte | |
| esoph- | gullet | esophagus | |
| eu- | true | eukaryote | |
| evolv- | unroll | evolution | |
| ex-/exo- | out of, outside | exothermic | |
| fasc- | bundle | fascicle | |
| fila- | thread | filament | |
| flacc- | flabby | flaccid | |
| gangli- | knot, gathered into a ball | ganglion | |
| gastr- | stomach, pouch | gastric juices | |
| gen- | birth, race | gene | |
| genesis | origin, creation | gametogenesis | |
| geri | to grow old | geriatrics | |
| glia- | glue | glial cell | |
| glomer- | to wind up | glomerulus | |
| glott- | tongue | epiglottis | |
| glut- | sticky | gluten | |
| glyc- | sweet, sugar | glucose | |
| -gram | record | cladogram | |
| -graph | recording | biogeography | |
| gymno- | naked | gymnosperm | |
| gyn- | woman | gynecology | |
| hal- | salty | halophile | |
| haplo- | single, simple | haploid | |
| helminth- | worm | Platyhelminthes | |
| hem- | blood | hemophilia | |
| hemi- | half | hemizygous | |
| hepat- | liver | hepatitis | |
| hetero- | different, mixed | heterozygous | |
| hibern- | to winter | hibernation | |
| hist- | causes to stand | histology | |
| holo- | whole | Holocene | |
| homo-/homeo- | alike, similar | homeostasis | |
| homo | man | *Homo sapiens* | |
| hormon- | that which excites | hormone | |
| hydr- | water | hydrate | |
| hyper- | above | hypertonic | |
| hypo- | below, beneath | hypothesis | |
| ichthy- | fish | Osteichthyes | |
| ile- | groin | ileum | |
| in- | in | innate | |
| incis- | cutting in | incisor | |
| inflexis- | rigid | flexion | |
| infra- | below, under | infraorder | |
| integument- | covering | integument | |
| inter- | between | interdependence | |
| intra- | within | intracellular | |
| -iole | diminutive | arteriole | |

| ROOT | MEANING | EXAMPLE | MORE EXAMPLES |
|---|---|---|---|
| iso- | equal | isotonic | |
| -itis | inflammation | arthritis | |
| karyo- | kernel, core | prokaryote | |
| kilo- | thousand | kilogram | |
| lact- | milk | lactation | |
| leuko- | white | leukocyte | |
| lign- | wood | lignin | |
| lip- | fat, grease | lipid | |
| -logist | one who studies | cytologist | |
| -logy | discourse, study of | microbiology | |
| lymph- | clear water | lymph vessel | |
| -lysis | to break, to loosen | hydrolysis | |
| macro- | large | macroscopic | |
| magnif- | splendid | magnification | |
| mal- | bad | malignant | |
| mamma- | breast | mammal | |
| mandib- | lower jaw | mandible | |
| mars- | pouch, purse | marsupial | |
| mater- | mother | maternal | |
| medi- | medium | medial | |
| meio- | to lessen | meiosis | |
| melano- | dark, black | melanin | |
| meno- | month | menopause | |
| meris- | division | meristem | |
| meso- | middle | mesoderm | |
| meta- | after, later | metaphase | |
| metabol- | change | metabolism | |
| micro- | small | microscopic | |
| mito- | thread | mitosis | |
| mix- | mingle | mixitroph | |
| mono- | one | monosomy | |
| morph | form, shape | morphology | |
| motor- | mover | motor neuron | |
| multi- | many | multipotent | |
| myco- | fungus | mycology | |
| myo- | muscle | myofibril | |
| nemato- | thread, hair | nematode | |
| nephro- | kidney | nephron | |
| neuro- | of the brain | neuron | |
| nomen- | name | nomenclature | |
| non- | not | nondisjunction | |
| noto- | back | notochord | |
| -oid | like | hominoid | |
| olfac- | smelling | olfactory | |
| oligo- | few | oligosaccharide | |
| -oma | tumor | carcinoma | |
| omni- | all | omnipotent | |
| onc- | tumor, mass | oncogene | |
| oo- | egg | oogenesis | |
| opt- | eye | optic nerve | |
| or- | mouth | oral cavity | |
| -osis | condition, process | meiosis | |
| osteo- | bone | osteocyte | |
| ov- | egg | oviduct | |
| pan- | all | pandemic | |
| para- | near, beside | paraphyletic | |
| patho- | disease | pathogen | |
| ped- | foot | biped | |
| pent- | five | pentose sugar | |

| --- | --- | --- | --- |
| pept- | to digest | polypeptide | |
| per- | large amount | peroxide | |
| period- | around, cycle | periodic table | |
| petro- | rock | petroleum | |
| phaeo- | dark | Phaeophyta | |
| phag | eat | phagocytosis | |
| pher- | to carry | pheromone | |
| -phil | love | hydrophilic | |
| phlo- | tree bark | phloem | |
| phor- | carry | electrophoresis | |
| photo- | light | photosynthesis | |
| phragm- | fence | diaphragm | |
| phyllo | leaf | chlorophyll | |
| phylo- | race, class | phylogenetic tree | |
| physio- | natural | physiology | |
| phyto- | plant | phytoplankton | |
| pil- | hair | pilus | |
| pinea- | pine cone | pineal gland | |
| placo- | flat surface | placoderm | |
| platy- | flat, broad | Platyhelminthes | |
| plum- | feather | plumage | |
| pneu(mo)- | air, lung | pneumonia | |
| pod | foot | pseudopod | |
| poly- | many | polydactyly | |
| post- | after | postzygotic | |
| pre- | before | prezygotic | |
| prim- | first | primer | |
| pro- | before, in front | prokaryote | |
| prom- | first | promoter | |
| protist- | first | Protista | |
| proxi- | nearest | proximal tubule | |
| pseudo- | false | pseudopod | |
| psycho- | mind | psychosomatic | |
| pter- | wing, fern | *Archaeopteryx* | |
| pube- | sexually mature, adult | puberty | |
| pulmo- | lung | pulmonary artery | |
| pyro- | heat, fire | Pyrrophyta | |
| quad/quat | four | quaternary consumer | |
| re- | back, again | reverse transcriptase | |
| rect- | straight | *Homo erectus* | |
| reflex- | bending back | reflex | |
| ren- | kidney | adrenal glands | |
| retina- | net-like | retina | |
| retro- | behind, backward | retrovirus | |
| rhin- | nose, snout | rhinoceros | |
| rhiz- | root | mycorrhizae | |
| rhodo- | red | Rhodophyta | |
| rode- | gnaw | rodent | |
| rostr- | beak | rostrum | |
| rumen- | throat | ruminate | |
| racchar- | sugar | polysaccharide | |
| sapro- | rotten | saprobe | |
| sarc- | comb | sarcomere | |
| saur- | lizard | dinosaur | |
| scala- | ladder | scala naturae | |
| schizo- | split | schizophrenia | |
| scler- | hard | atherosclerosis | |
| scop- | gaze, shadow | microscopy | |
| seba- | tallow, wax | sebaceous gland | |

| ROOT | MEANING | EXAMPLE | MORE EXAMPLES |
|---|---|---|---|
| semi- | half | semipermeable | |
| sept- | wall | septum | |
| ser- | body fluid | serum | |
| set- | hair | setae | |
| simi- | monkey | simian | |
| sol- | sun | solar radiation | |
| soma- | body | chromosome | |
| specie- | sort, kind | species | |
| sperm- | seed | spermatogenesis | |
| sphinct- | closing | sphincter muscle | |
| squam- | scale | squamous | |
| stoma- | mouth | stomata | |
| strat- | thing spread out | stratum | |
| strept- | twisted | streptococcus | |
| strict- | upright, stiff | constrict | |
| styl- | column, pillar | style | |
| sub- | below, under | subspecies | |
| sucr- | sugar | sucrose | |
| super- | above | superposition | |
| sutur- | seam | suture | |
| sym- | with | symbiosis | |
| syn- | joined together | dehydration synthesis | |
| system- | organized whole | body system | |
| tact- | touch | tactile | |
| tele- | at a distance, end | telomere | |
| terti- | third | tertiary consumer | |
| therm- | heat | thermal energy | |
| -thesis | to put, to place | hypothesis | |
| -tome | slice | appendectomy | |
| -ton | tone | hypertonic | |
| topo- | place | isotope | |
| toti- | whole | totipotent | |
| toxi- | poison | biotoxin | |
| trache- | windpipe | trachea | |
| trans- | across | transmembrane protein | |
| tri- | three | triploidy | |
| trop- | turning | troponin | |
| -troph | food | trophic level | |
| uni- | one | uniformitarianism | |
| uro- | tail | urogenital | |
| vac- | cow | vaccine | |
| vagina- | sheath | vagina | |
| vas- | vessel | vascular plant | |
| ven- | vein | vena cava | |
| ventr- | belly | ventral | |
| visc- | organs of body cavity | visceral | |
| vita- | life | vital force | |
| vora- | devour | omnivore | |
| xantho- | yellow | xanthophyll | |
| xyl- | wood | xylem | |
| zo- | animal | protozoan | |
| zyg- | yoke | zygote | |

# Answer Key

**1·1**
1. An organism is any living thing.
2. Homeostasis is the maintenance of stable internal conditions when external conditions are changing. (*homeo* = same)
3. Metabolism is the sum of all the chemical reactions within a cell and/or organism, including digestion/decomposition and synthesis. (*metabole* = change)
4. Observation is part of the scientific method and entails utilizing one or more of the five senses to perceive natural events.
5. A controlled experiment compares a control group to an experimental group in an attempt to support or reject a hypothesis.

**1·2**
1. d   2. b   3. a   4. c   5. b

**1·3**
1. Mold is made up of individual cells that can perform homeostasis, metabolize the substrate it lives upon, and grow through cell enlargement and cell division. Accumulating ice forms as water molecules move from the liquid to solid state in response to temperature. The ice is not cellular but instead is composed of water molecules that crystallize as hydrogen bonds form.
2. If no control group is present, then there is no point of comparison for the experimental group. Any results may or may not be due to the independent variable being tested; there is no way to know for sure.
3. Magnification allows an increase in the apparent size of an object, while resolution allows for a clear, crisp image of the specimen being observed.

**1·4**
1. D   2. C   3. A   4. B

**1·5**
1. *Regulation and feedback* is demonstrated through the notion of homeostasis, one of the characteristics of life. An organism can maintain appropriate internal conditions in spite of a changing external environment.
2. *Science methodologies and applications to society* is demonstrated through the microscope. A microscopist uses the tool to investigate microorganisms that would otherwise go unknown.
3. *Natural interdependence* is demonstrated through the notion of a multicellular organism. The cells within are interdependent and must communicate to keep the entire organism functioning properly.

**2·1**
1. An atom is the smallest naturally occurring unit of matter; a molecule is made up of two or more atoms covalently bonded together.
2. An element is a pure substance, characterized by a given number of protons in the atomic nucleus; a compound is formed by two or more different elements that are bonded together.
3. Polarity describes the uneven sharing of electrons between atoms in a covalent bond; a hydrogen bond can result between the partial charges that result from a polar covalent bond.

**2·2**
1. d   2. b   3. a   4. a   5. d

**2·3**
1. Oxygen gas constitutes a molecule but not a compound. It is made up of two atoms covalently bonded together, but the two atoms are the same, so it is still a pure element.
2. In forming a large drop of water, hydrogen bonds between water molecules contribute to cohesion. In holding the drop to the faucet, hydrogen bonds contribute to adhesion.

3. Because solid water is less dense than liquid water, a small pond will not freeze throughout but will instead produce a layer of floating ice on the surface. Organisms can thus survive in the liquid water below until the melt.

**2·4**  1. C  2. B  3. A  4. This reaction is endothermic, as evidenced by the products existing at a higher energy state than the reactants.

**2·5**  1. *Form facilitates function* is demonstrated through the polarity of water and the characteristics that result. Because of an uneven distribution of charges throughout the water molecule, hydrogen bonds can form, making water sticky and a good solvent.

2. *Energy and organization* is demonstrated through a chemical reaction. As energy is transferred, atoms are rearranged as chemical bonds break and re-form.

3. *Science methodologies and applications to society* is demonstrated through the periodic table of elements. This tool allows chemists to organize all of the elements in a methodical way such that information like atomic number and atomic mass are easily accessible.

**3·1**  1. A monomer is the fundamental building block of biological macromolecules like proteins, carbohydrates, and nucleic acids. When monomers are bonded together, a polymer is created.

2. A dehydration synthesis is the means by which two monomers are bonded together to form a polymer. A polymer can be broken down into individual monomers through hydrolysis reactions.

3. A nucleic acid is one type of biological macromolecule that is used to store and express the genetic code. Polymer forms like DNA and RNA comprise individual monomer units called *nucleotides*.

**3·2**  1. b  2. d  3. c  4. c  5. a

**3·3**  1. Each carbon atom possesses four valence electrons, so it can form up to four (single) bonds with other atoms. This allows for straight chains, branched chains, and rings to form between carbon atoms typical in biological macromolecules.

2. The twenty-six letters in the English alphabet are like the twenty amino acids that exist in nature. Any number of different letters can be arranged together in a specific sequence to make up a word, which would be analogous to a polymer.

3. Waxes secreted on the outer surface of an organism are usually used for waterproofing. By creating a watertight exterior, an organism has more control over the quantity of water it loses or absorbs.

**3·4**  1. Phosphate group  2. Nitrogenous base  3. Pentose sugar

**3·5**  1. *Continuity and change* is demonstrated by the metabolic reactions of dehydration synthesis and hydrolysis. Monomers are attached to one another to make polymers, and polymers are digested into individual monomer components as needed by the organism.

2. *Energy and organization* is demonstrated by lipids. Relative to carbohydrates, lipids, like triglycerides, are nonpolar and hydrophobic and are much harder to break down for use as cellular energy. Triglycerides composed of saturated fatty acid chains are especially challenging because of the very stable, single-bond structure within the carbon chain.

3. *Form facilitates function* is demonstrated by the various forms of carbohydrates. When in disaccharide form, only one bond needs to be broken to release the monosaccharides, so they are available as quick energy. When in polysaccharide form, they are considered a longer-term energy source because of the multiple bonds that must be broken to release all of the monosaccharides for energy.

**4·1**  1. The cell membrane is the boundary just outside of the cell's cytoplasm. Its phospholipid bilayer structure makes the membrane selectively permeable, able to regulate what enters and leaves the cell.

2. Osmosis is the diffusion of water molecules, the movement of those molecules down the concentration gradient from a region of relatively higher concentration to a region of lower concentration.

3. A vesicle is a small, membrane-bound container made during the process of phagocytosis when a cell engulfs a solid particle (or an entire cell) detected outside of its membrane.

**4·2**  1. c  2. a  3. d  4. b  5. d

**4·3**  1. The cell theory states the following: all living things are composed of at least one cell; the cell is the basic unit of structure and function for life; and all cells come from preexisting cells.

2. Eukaryotic cells possess membrane-bound organelles like the nucleus or the mitochondrion, while prokaryotes do not. The significance of possessing such organelles is that the membrane allows for each organelle to maintain its own structure and function while at the same time allows for integration with other organelles within the same cell.

3. If Laura were to drink the salt water, the salt concentration of her bloodstream would increase after ingestion and absorption. Normally, her red blood cells are isotonic (having the same osmotic pressure)

with the bloodstream, but now her red blood cells would lose water through osmosis due to the concentration gradient.

**4·4**   1. Chloroplast   2. Central vacuole   3. Cell wall   4. Nucleus

**4·5**   1. *Form facilitates function* is demonstrated through the structure of the cell membrane. The phospholipids in the bilayer are oriented such that their hydrophilic heads point outward toward the aqueous external environment and inward toward the aqueous cytosol of the cell. The hydrophobic tails are left to point toward the interior of the membrane, creating an environment that prevents most polar or ionic substances from passing.

2. *Regulation and feedback* is demonstrated through the process of active transport. If a cell requires an imbalance of ions to function properly, as observed in the sodium-potassium pump of nerve cells, then it must go against equilibrium to maintain the proper concentration gradient. The cell invests energy in the form of ATP in order to do so.

3. *Continuity and change* is demonstrated through the process of exocytosis. A waste vesicle moves toward the cell membrane and eventually fuses with the membrane in order to release its contents into the external environment. The phospholipids that made up the membrane of the vesicle are still part of the cell, but they have been relocated to the outer plasma membrane.

**5·1**   1. An enzyme is a biological catalyst composed of protein. Its function is to lower the activation energy required for reactants to form products, acting to speed up chemical reactions. It does so by arranging reactants in the optimal energy conformation at the enzyme's active site. After the reaction reaches completion, the products are released from the active site, and the enzyme is ready to act again.

2. Pigments like chlorophyll are concentrated in the chloroplasts, allowing those organelles to trap light energy and redirect that energy into substances like NADPH and ATP that are then used to build glucose for food (chemical energy).

3. The mitochondrial matrix is the region inside the organelle that surrounds all of the cristae. Once the citric acid cycle has been carried out there, the products (NADH and $FADH_2$) are available for the processes of the electron transport chain and chemiosmosis, the final steps in aerobic cell respiration.

**5·2**   1. b   2. c   3. d   4. a   5. b

**5·3**   1. Enzymes function as biological catalysts, substances that speed up chemical reactions, by lowering the activation energy required for the reactants to form products.

2. Both the chloroplast and the mitochondrion are well suited for energy transfer processes because they both have increased internal surface area. This comes in the form of thylakoids in the chloroplast for maximum light absorption and in cristae in the mitochondrion for maximum electron transport chain and chemiosmotic processes.

3. By continuing lactic acid fermentation while in oxygen debt, muscle cells can still generate some ATP through the breakdown of glucose, although it is much less efficient at only two ATPs per glucose. If this continues for a prolonged period or to an extreme extent, then the person's muscles may run out of energy and cramp up, or the person may faint.

**5·4**   1. B   2. C   3. D   4. F

**5·5**   1. *Form facilitates function* is illustrated by the structure of the chloroplast. It contains pigments that are specially structured to capture certain light waves. They then direct that energy into the building of glucose through the coordination of the light-dependent and light-independent reactions within the thylakoids and stroma of the chloroplast.

2. *Energy and organization* is illustrated by the dual metabolic processes of catabolism and anabolism that cells and organisms carry out. The polymers that an animal has ingested are catabolically broken down into monomers. The monomers are then available for use to build up new polymers that are needed by that animal. To maintain the overall structure of an organism, energy is constantly invested to work against entropy.

3. *Natural interdependence* is illustrated by the autotrophs that conduct photosynthesis and the heterotrophs that rely solely on cell respiration metabolically. The autotrophs need the carbon dioxide given off by the heterotrophs during cell respiration, and the heterotrophs need the oxygen given off as a photosynthetic by-product.

**6·1**   1. An autosome is a non-sex chromosome. (*auto* = self; *soma* = body)

2. A diploid cell is one that has two sets of DNA arranged in homologous pairs of chromosomes. (*diplo* = twofold; *oid* = appearance of)

3. Interphase is the time between cell divisions during which the cell carries out DNA replication and protein synthesis. (*inter* = between; *phase* = a stage of)

4. Cytokinesis involves the splitting of the cytoplasm and the formation of separate cell membranes that usually coincides with the end of mitosis. (*cyto* = cell; *kinesis* = movement)

5. Oogenesis is the specialized type of meiosis that results in the formation of the ovum, the female gamete. (*oo* = egg; *genesis* = creation)

**6·2**　1. b　2. a　3. c　4. d　5. c

**6·3**　1. Cytokinesis in plant cells involves the deposition of a cell plate down the midline of the original cell until the new cell wall is finally created. In animal cells, cytokinesis involves the formation of a cleavage furrow as the new membranes form between offspring cells.

2. Mitosis without cytokinesis would produce a multinucleate cell—one with many nuclei. This is actually the typical structure of skeletal muscle cells in vertebrates. The multiple nuclei help to support the very active metabolism of muscle cells.

3. Oogenesis produces four haploid cells just as spermatogenesis does, but only one mature egg results. This allows the egg to contain most of the cytoplasm from the original cell, thus it is best prepared for fertilization and eventual growth of an embryo. Spermatogenesis produces four equally sized sperm, all small and motile to increase the chances of reaching the egg.

**6·4**　1. Metaphase　2. Telophase　3. Prophase　4. Anaphase

**6·5**　1. *Regulation and feedback* is demonstrated through the cell cycle. Only if the cell receives certain environmental cues does it move from the $G_1$ phase into the S phase. If the cell is not certain that it will ever divide, it will not enter the S phase during which it expands energy and other resources to replicate its DNA but will instead be held in the $G_0$ phase permanently or until environmental conditions change.

2. *Energy and organization* is demonstrated through chromosome packing. In order to fit all six feet of DNA that humans have in each of their cells, the DNA double helix is condensed, as it is wound tightly around histones into chromatin. When the DNA needs to be even more highly organized during mitosis (or meiosis), it further packs the chromatin into chromosomes.

3. *Continuity and change* is demonstrated through the phases of meiosis. Meiosis I carefully coordinates the separation of homologous pairs of chromosomes to create two haploid cells. These haploid cells then undergo the sequenced events of meiosis II to pull each chromosome apart into chromatids and allocate those chromatids to separate cells.

**7·1**　1. A mutation is any change in the sequence of nucleotides within a DNA molecule. (*mutatio* = change)

2. DNA replication is the means by which a cell copies all of the DNA in its genome in preparation for an eventual round of mitosis. (*replica* = repeat)

3. RNA polymerase is the enzyme that catalyzes the formation of the RNA transcript during transcription. (*poly* = many)

4. Transcription is the process by which a DNA gene is read and used to make a complementary RNA copy. This RNA transcript can then be utilized to synthesize a protein. (*trans* = across; *script* = to write)

5. A genome is the entirety of the genes carried on all of the chromosomes typical for a given species of organism. (*gen* = birth, race)

**7·2**　1. c　2. c　3. d　4. b　5. a

**7·3**　1. Within any DNA molecule, strong covalent bonds hold all of the parts of any single nucleotide together and hold adjacent nucleotides to one another within each strand. This ensures that the specific sequence of nucleotides along one strand (and within a gene) stay constant. Weaker hydrogen bonds are found only between the two strands of nucleotides in the double helix, allowing for times when the strands can be separated and complementary copies of their sequences made.

2. Replication would produce A – T – G – T – C – C – C – A – T – A – T – T – G – A – C – T – A – G. Transcription would instead produce A – U – G – U – C – C – C – A – U – A – U – U – G – A – C – U – A – G. The sequence of tRNA anticodons that would be involved in translation is U – A – C – A – G – G – G – U – A –U – A – A – C – U – G – A – U – C.

3. DNA needs to be double stranded in order to protect the meaningful code on one strand with the complementary sequence of nucleotides on the other strand. RNA must be single-stranded in order to work effectively, for the sequence of bases needs to be easily accessible during translation.

**7·4**　1. DNA replication　2. Transcription　3. Translation

**7·5**　1. *Form facilitates function* is demonstrated by the structure of tRNA. One end of the molecule contains the anticodon, the sequence of three nucleotides complementary to a codon on mRNA. This tRNA carries the appropriate amino acid indicated by that mRNA codon on the opposite end of the tRNA molecule.

2. *Energy and organization* is demonstrated by the role of helicase during DNA replication. The enzyme transfers enough energy to the DNA molecule to unwind a portion of the double helix and break the hydrogen bonds holding the two strands together.

3. *Continuity and change* is demonstrated by the notion of the central dogma. The genetic code is stored in the form of DNA, but can be expressed as proteins through the action of various forms of RNA. If a mutation occurs and the original DNA sequence is changed, then the resultant protein produced from translation of that sequence would also likely change.

**8·1**
1. Genetics is the field of biology that studies heredity, the transmission of genetic traits from parents to offspring.

2. Genotype describes the specific combination of alleles that an individual possesses for a given gene; phenotype describes the physical expression of that genotype.

3. Dominant is used to describe an allele that masks the expression of the other, recessive allele for the given gene. If at least one dominant allele is present, then the dominant phenotype is expressed.

**8·2**
1. b  2. d  3. c  4. a  5. b

**8·3**
1. The process of the separation of homologous pairs that occurs during meiosis I demonstrates Mendel's law of segregation. Thus only one allele from the individual's genotype will actually be transmitted to a potential offspring through a gamete. Mendel's law of independent assortment is also observed in meiosis I through the random alignment of chromosomes along the equator. Each chromosome pair operates independently of the others, so the genes carried are inherited independently of the genes carried on a different pair.

2. A roan cow would not make a good subject for a testcross. For a testcross, there should be one individual with a dominant phenotype but unknown genotype, and that individual is mated with another individual that has the recessive phenotype/genotype. A roan cow is a red and white cow that results from codominant coat color alleles, so this type of cow would not be suitable for a testcross genetically.

3. Ten offspring should be pink (genotype = *RW*), and ten offspring should be white (*WW*).

**8·4**
1. Terminal genotype = *aa*
2. 1:1 (1 *Aa* for every 1 *aa*)
3. 1:1 (1 axial for every 1 terminal)
4. 50

|  | A | a |
|---|---|---|
| a | Aa | aa |
| a | Aa | aa |

**8·5**
1. *Science methodologies and applications to society* is represented through the use of Punnett squares to predict genetic outcomes in successive generations of offspring. This is useful in agriculture, animal breeding, and in understanding our own transmission of traits as humans.

2. *Continuity and change* is represented through heredity itself. As a sexually reproducing organism creates gametes, it only packages half of its DNA (a mix of one of the two sets of information it inherited from its parents) to package in each gamete. That way, when gametes fertilize, a unique combination of genes is created in the new individual.

3. *Natural interdependence* is demonstrated through the pollination process that Mendel mimicked in his experiments. The eggs that a plant produces wait in the flower until pollen arrives there carrying sperm. Only then can gametes merge, a zygote form, and the DNA carried in the gametes combine and begin coding for a new organism.

**9·1**
1. A sex-linked trait houses its genes on one of the chromosomes that determines biological sex. In that way, its expression is linked to the expression of biological sex itself.

2. A point mutation is a change in a single nucleotide within a gene. It can take the form of a substitution, insertion, or deletion. (*mut* = change)

3. A polygenic trait is one that is dependent on the genetic instructions carried by several individual genes. (*poly* = many; *gen* = race)

4. A DNA fingerprint is a means of establishing genetic identity using the DNA from an individual that creates unique banding patterns once digested.

5. Restriction enzymes are naturally occurring in bacteria but are used by humans in many biotechnologies. They act by cutting DNA open, often leaving sticky ends that allow DNA strands to be easily manipulated and bioengineered.

**9·2**    1. c   2. d   3. d   4. b   5. a

**9·3**    1. A substitution mutation may or may not actually result in a change in the resultant protein. Because of the redundancy built into the code (i.e., more than one DNA triplet codes for the same amino acid), then a substitution mutation may in fact change a DNA triplet without affecting the corresponding amino acid or the resultant protein.

         2. A sex-linked trait has its genes on a sex chromosome and is thus directly linked to the expression of biological sex itself. A sex-influenced trait has its genes housed on an autosome, but the expression of the resultant protein is influenced by the presence of sex hormones.

         3. Producing a DNA fingerprint involves obtaining DNA samples from the subjects of interest, digesting the DNA samples using restriction enzymes, using gel electrophoresis to separate the fragments of DNA created for the digestion, and then analyzing the banding patterns for matches.

**9·4**    1. Mother's phenotype = type B; father's phenotype = type A

         2. Possible offspring genotypes = $I^A i$ (AO), $I^B i$ (BO), $I^A I^B$ (AB), and $ii$ (OO)

         3. 3/4

         4. 1/4

**9·5**    1. *Natural interdependence* is demonstrated through the concept of nondisjunction. Normally, two homologous chromosomes are present in a new zygote, one chromosome being donated from each parent via egg and sperm. When nondisjunction occurs, one of the gametes is packaged with one extra/missing chromosome, thus one "pair" of homologues in the zygote is not really a pair at all; it is instead a solo chromosome or a triplet, and in either case, this presents problems for the individual.

         2. *Science methodologies and applications to society* is exemplified by Morgan's work with the genetics of fruit flies. Through careful observations and analyses, his team was able to decipher the chromosomal pattern for biological sex and was able to uncover the mechanism behind sex-linked traits.

         3. *Form facilitates function* is shown through the action of restriction enzymes in gene cloning. Because a restriction enzyme naturally creates sticky ends when it cuts through DNA, the DNA is capable of being spliced to other segments of DNA through the re-formation of hydrogen bonds between complementary base pairs.

**10·1**    1. Spontaneous generation was the old notion that simple forms of life could spontaneously take form without the influence of life and possible due to some "vital force." Biogenesis is the accepted notion that only current life can generate new life.

         2. Relative age describes whether a fossil is older or younger than another, while absolute age is the actual age of the fossil in years.

         3. A homologous structure is one that is derived from a common ancestor and found in relatively closely related species. Often homologous structures are identifiable among species that demonstrate divergent evolution, a pattern in which one ancestral species separates into two separate species.

         4. Evolution is the change in allele frequencies in a population over time. Natural selection, the notion that individuals with best-fit traits disproportionately survive and reproduce in a population, is the mechanism by which evolution occurs.

         5. Gradualism describes the rate of speciation in which a slow, gradual phenotypic change occurs to create a new species over a very long period of time. Punctuated equilibrium describes the relatively sudden appearance of new species followed by long periods of no relative change.

**10·2**    1. a   2. b   3. c   4. b   5. c

**10·3**    1. The law of superposition helps to establish the relative age of a fossil by comparing fossils discovered in one stratum with those found in strata immediately above or below.

         2. For a bacterium to evolve the ability to metabolize plastic, a bacterium must first experience a mutation that changes its original genetic instructions. If the protein created does confer this new trait, and if the trait is favorable in the environment, then the individual possessing this new gene will be selected for in the population. All of the asexual offspring produced from this bacterium will also possess this trait, so very quickly this population may evolve to show a high frequency (from an initial frequency of zero) of this trait.

         3. Microevolution involves the smallest scope of evolutionary change, namely the change in the frequency of an allele in a population over time. Macroevolution is the largest scope of evolutionary change, specifically the creation of a new species from an ancestral one (speciation).

**10·4**    1. Graph 2

         2. Graph 1

         3. Species A and B

         4. Species C and D

**10·5**
1. *Continuity and change* is represented through the very notion of evolution itself. If a mutation (change) occurs in DNA, the new trait conferred may present an evolutionary advantage or disadvantage. Over time, this helps the species adapt to its environment.
2. *Science methodologies and applications to society* is demonstrated through the act of absolute dating. Using radioisotopes, the absolute age of a fossil in years can be determined, and the evolutionary history of that lineage can be better understood.
3. *Regulation and feedback* is shown through the concept of the Hardy-Weinberg equilibrium. If the environment is relatively stable, then there is likely little pressure on the populations of organisms living within it to evolve. If some environmental condition changes, however, then the equilibriums shift, and the populations respond through evolution and adaptation.

**11·1**
1. Binomial nomenclature is the system developed by Linnaeus for naming organisms that relies on both the genus and species names. (*bi-* = two; *nomen* = name)
2. Taxonomy is the field of biology that studies classification and devises new classification schemes. (*taxon-* = level)
3. A phylogenetic tree is a graphical means of demonstrating the evolutionary relationships between organisms in a select group. (*phylo-* = class and *gene* = race)

**11·2**
1. d  2. b  3. d  4. a  5. b

**11·3**
1. Taxonomy is a constantly changing science. The number of unique species known to science is always changing, and occasionally new technologies present new evidence that forces taxonomists to reconsider previous classification schemes.
2. Domain Eukarya is distinguishable from the other domains of life because its members possess a nucleus within each cell. Archaea can be distinguished from Eubacteria in that its members have genes that contain introns.
3. Criteria currently used to classify organisms that would not have been utilized by Aristotle or Linnaeus include genetic homologies revealed by comparative biochemistry techniques like DNA hybridization and protein sequencing.

**11·4**
1. Species 6
2. Species 2 and 3
3. Species 5
4. Species 1

**11·5**
1. *Form facilitates function* is demonstrated through the structure of a phylogenetic tree or cladogram. The base of the tree (root) represents the common ancestor, the branches represent speciation events over time, and the final tips (buds) represent the various unique species that resulted.
2. *Continuity and change* is shown through the notion of evolution. As the frequency of an allele or a trait changes over time, the population is evolving.
3. *Science methodologies and applications to society* is exemplified by the field of taxonomy itself. Taxonomy is used by humans to organize, categorize, and keep track of the millions of unique species known to science.

**12·1**
1. Acellular refers to a structure that is not composed of even one cell. Viruses have this characteristic, although they have some cell-like structures. (*a-* = not)
2. A bacteriophage is a virus that infects prokaryotic cells. (*phag-* = eat)
3. A provirus is a segment of viral DNA inserted into the main bacterial chromosome of the infected cell. It will later be excised and used to make new virus. (*pro-* = before)
4. A vaccine is a substance that, when exposed to the immune cells of the blood, will prompt the production of antibodies that will fight off that particular virus. The first vaccine utilized cowpox-infected serum in an attempt to inoculate humans from smallpox. (*vac-* = cow)

**12·2**
1. c  2. b  3. a  4. d  5. a

**12·3**
1. The lytic cycle involves the immediate transcription and translation of new virus once the bacteriophage has inserted its DNA into the host cell. The lysogenic cycle instead involves a latent phase, for the bacteriophage DNA is inserted in the prokaryote's main chromosome where it remains until conditions are optimal for infection. Then the prophage DNA is excised out of the chromosome, and the cell enters the lytic cycle, during which virus is actively made.
2. A virus causes infection in a eukaryotic cell by interacting with the host cell's receptors and tricking the cell into performing endocytosis. Once inside the cytoplasm, the viral DNA is incorporated into a chromosome of the host cell where it can be used to make new virus through protein synthesis. An example of a DNA virus that infects humans is the rabies virus; an RNA virus that infects humans is influenza.

3. A vaccine is useful in preventing a viral infection because it boosts the immune system to make antibodies against that particular virus before the body ever actually is exposed to it. If an infection has already occurred, an antiviral drug may instead be administered to try to prevent or lessen further replication of the virus.

**12·4**   1. Envelope   2. Capsid   3. Nucleic acid

**12·5**   1. *Energy and organization* is demonstrated by the classification of viruses outside of the realm of life. Given that viruses lack cellular organization and the ability to metabolize, they do not meet two very basic characteristics of life.

2. *Form facilitates function* is shown through the structure of a retrovirus. Because a retrovirus is packaged with RNA instead of DNA, it also must come packaged with the reverse transcriptase enzyme, one that catalyzes the production of viral DNA from viral RNA.

3. *Continuity and change* is represented through the lytic and lysogenic cycles of bacteriophages. After infection of a prokaryote, a bacteriophage may first enter the lysogenic cycle during which it stores its DNA in the bacterium's chromosome and is copied during asexual reproduction. When conditions are optimal, the viral DNA will be excised as the virus enters the lytic phase. Active virus is made, and the cell will eventually burst.

**13·1**   1. An extremophile is an organism that prefers conditions that most other organisms would find severe or extreme. Archaebacteria are extremophiles, preferring very salty, very hot, or oxygen-free environments. (*-phil* = loving)

2. A Gram-positive bacterium is one that stains dark blue/purple when treated with the Gram stain and viewed under the microscope. An example is *Streptococcus*.

3. An antibiotic is a substance that kills bacterial cells. Many fungi and plants produce natural antibiotics. The first antibiotic known was penicillin. (*anti-* = against; *bio-* = life)

**13·2**   1. c   2. b   3. b   4. d   5. a

**13·3**   1. Archaebacteria is more closely related to Eukarya than to Eubacteria, as evidenced by the presence of introns in their genes and the structure of their rRNA and ribosomes.

2. Humans contribute to antibiotic resistance by overusing and misusing antibiotics and antibacterial products. This only encourages resistance in the population by allowing only those bacteria that possess the resistance trait to survive and reproduce.

3. More than 99 percent of the bacteria on the planet are not pathogenic, so the vast majority are either helpful to humans or are neutral in their effects. Many are used in food processing and in the production of pharmaceutical products, and countless inhabit our bodies contributing to metabolism and immunity.

**13·4**   1. Pilus   2. Plasmid   3. Ribosome   4. Cell wall

**13·5**   1. *Continuity and change* is represented by the development of antibiotic resistance in certain strains of bacteria. As a bacterium encounters another in the environment, it may copy and exchange a plasmid containing resistance genes through a pilus. This can confer the trait to the receiving bacterium.

2. *Form facilitates function* is demonstrated through the structure of the glycocalyx. This fuzzy carbohydrate coat helps some bacteria attach more readily to their host cell.

3. *Science methodologies and applications to society* is exemplified through the notion of bioremediation. Microbiologists study natural metabolic action of different prokaryotes and then utilize those that are specially adapted to specific environments (e.g., those that can metabolize oil) to help treat problems in the environment.

**14·1**   1. Pseudopodia are the false feet formed by a protozoan through the process of cytoplasmic streaming. This occurs when cytoplasm is forced to one extension of the membrane, creating a pseudopodium.

2. An algal bloom is a population explosion of algae. One specific type called a *red tide* is potentially harmful for other organisms living in the water or consuming those organisms and/or the water itself due to high concentrations of an algal biotoxin.

3. Slime molds are fungus-like protists that tend to be decomposers and are characterized by a unique collective phase. Water molds are also fungus-like protists, but they instead tend to be unicellular parasites of fish and other aquatic life.

**14·2**   1. c   2. d   3. d   4. d   5. a

**14·3**   1. The diversity of kingdom Protista is represented by morphology (i.e., there are both unicellular and multicellular varieties), ecological role (i.e., there are producers, consumers, and decomposers), and reproductive type (i.e., there are asexual and sexual types).

2. Because of the extreme diversity observed within the kingdom, protists are often classified as plant-, fungus-, or animal-like. These are not true phylogenetic groups, as they are not based necessarily on evolutionary history but are based instead on the ecological role played by the fungus.

3. Sushi, ice cream, and salad dressing are commonly consumed protist food products.

**14·4**  1. Animal-like protist   2. Protozoa   3. Heterotrophic

**14·5**  1. *Science methodologies and applications to society* is represented through the use of certain protists, the diatoms, in filters for pools and aquaria. These diatoms naturally filter the water as part of their metabolism, so this function is captured and used by humans as "diatomaceous earth."

2. *Form facilitates function* is demonstrated by the structure of algae. All algal groups possess a unique collection of photosynthetic pigments that not only produce that group's characteristic color but also often dictate the specific depths at which it can grow.

3. *Continuity and change* is exemplified by the life cycle of a typical slime mold. While much of the time the slime mold is unicellular, there is another phase of its life cycle during which it is joined together with other slime molds to create a macroscopic colonial slime mold for reproduction and feeding purposes.

**15·1**  1. A saprobe is a decomposer, an organism that relies on the consumption of dead and decaying organic matter for food. (*sapro-* = rotten)

2. A mycelium is the collective structure composed of individual filamentous hyphae observed in multicellular fungi. (*myco* = fungus)

3. A mycorrhizae is a mutually beneficial symbiotic relationship between a fungus and the root of a plant. (*myco* = fungus; *rhiz* = root)

**15·2**  1. b   2. d   3. b   4. d   5. c

**15·3**  1. Heterotrophy in fungi can be described as absorptive (i.e., they secrete digestive enzymes into the soil and then absorb the already-digested food). Heterotrophy in animals is described as ingestive (i.e., they ingest large pieces of organic food and then digest it internally before absorption).

2. Fungi were commonly misclassified as plants before their metabolism and morphology were well understood. This is understandable given that fungi typically grow in the soil like plants and have somewhat of a similar morphological structure. They are more closely related to animals, however, as evidenced by their heterotrophic lifestyle and their ability to store excess sugar as glycogen.

3. One positive interaction between humans and fungi is demonstrated in the production of antibiotics based on the natural fungal antibiotic penicillin. One negative interaction is observed in a human infected with athlete's foot, a common fungal skin infection.

**15·4**  1. Coenocytic   2. Club fungus   3. Mushroom

**15·5**  1. *Energy and organization* is demonstrated through the unique heterotrophic pattern observed in fungi. Their morphological organization allows them to secrete digestive enzymes into the soil where digestion of decaying organic matter occurs. They then absorb the simpler monomers and use them for energy.

2. *Natural interdependence* is exemplified by the symbiotic relationship between a fungus and an alga collectively called a lichen. The fungus gains sugars produced by the alga through photosynthesis, while the alga, living just under the surface of the fungus, gains protection from dehydration and other environmental threats.

3. *Science methodologies and applications to society* is represented by the production of antibiotics from natural fungal products. These and other synthetic antibiotics have saved countless lives of those infected by pathogenic bacteria.

**16·1**  1. A nonvascular plant is one that lacks any internal tubes for movement of water and dissolved solutes; this severely limits the overall body size of the plant. A vascular plant is one that evolved after the adaptation of vascular tissue, so these plants in many cases achieve impressive extremes in height.

2. Spores are haploid reproductive structures produced from the sporophyte. Through mitosis a spore can germinate and become a multicellular gametophyte plant. Seeds are dormant plant embryos, produced through mitosis of a plant zygote, that await proper environmental cues before germinating and producing the adult sporophyte plant.

3. Gymnosperms are vascular plants that possess uncovered seeds, sometimes on cones. Angiosperms also possess seeds, but they cover their seeds within a flower and often also eventually within a fruit.

**16·2**  1. d   2. c   3. a   4. b   5. c   6. d   7. a

**16·3**  1. Mosses possess waxy cuticles to prevent excess water loss on land. Ferns possess vascular tissue, internal tubes for water movement. Gymnosperms possess naked seeds often presented on cones. These dormant plant embryos inside a protective coat will eventually be dispersed, germinate, and may develop into the next generation gymnosperm. Angiosperms have covered seeds in the form of flowers and sometimes fruits. Flowers protect the developing seed deep within the ovule until ready for dispersal; when fruits form, they encourage dispersal of the seeds by enticing animals to eat them.

2. Plants also need inorganic nutrients from the soil. If the soil has not been changed or the plant has not been "fed" with replacement nutrients, then the plant may be malnourished.

3. Plants function in many significant ecological roles. For example, they are ecosystem producers, making glucose for the rest of the food chain. They also absorb carbon dioxide and thus participate in the carbon cycle. All paper and wood products are derived from plants, as are many pharmaceutical and personal care products.

**16·4**   1. Meiosis   2. Germination   3. Fertilization   4. Mitosis

**16·5**   1. *Regulation and feedback* is represented through the action of plant hormones like auxin. If a plant detects that one side of its body is not receiving adequate sunlight, the hormone targets those cells and encourages elongation. The result is that the plant is able to slowly reorient itself toward the sun.

2. *Natural interdependence* is demonstrated through the relationship of flowering plants and their animal pollinators. The relationship is mutually beneficial; the plant experiences increased reproductive success, while the pollinator receives a meal in the form of nectar and/or pollen itself.

3. *Continuity and change* is exemplified through the alternation of generations life cycle. When the plant in question is in the gametophyte generation, it is haploid and makes gametes to create the next sporophyte generation after fertilization. The sporophyte generation of the plant is diploid and produces haploid spores. Once released, each spore can germinate into a new sporophyte plant.

**17·1**   1. *Sessile* describes animals like sponges that do not have a means of purposefully moving about in their environment. They instead are usually attached to a substrate. Motile animals instead have a means of directing movement in their environment, like insects and birds.

2. *Indirect development* describes a developmental pattern in which the initial immature form and the final adult form look very different in overall morphology, as seen in insect metamorphosis. In direct development, as observed in placental mammals, the young is born already looking somewhat like an immature form of the adult organism.

3. Invertebrates are those animals that lack a vertebral column (or backbone), like worms, mollusks, and cnidarians. Vertebrates are animals that possess a vertebral column to protect the spinal cord, like fishes, reptiles, and mammals.

4. Oviparous animals, like reptiles and monotremes, produce eggs that hatch live young. Viviparous animals give birth to live young, like many members of Chondrichthyes and all marsupials and placental mammals.

5. An endotherm is an animal that has physiological control over its internal body temperature, like birds and mammals. Ectotherms are instead those animals that conform to the temperature of the surroundings and must rely on behavioral means to adjust when outside of the typical temperature range.

**17·2**   1. c   2. a   3. c   4. a   5. a   6. b   7. d

**17·3**   1. Vertebrates are distinguished from invertebrates in that they possess a vertebral column to protect the dorsal nerve cord. The vertebrae are part of the endoskeleton, which may comprise either cartilage or bone. Vertebrates also have pharyngeal gill slits at some point during development. The protection of the nerve cord and the development of gill slits were both evolutionarily fit traits that conferred a selective advantage to those organisms possessing them.

2. Those animals with bilateral symmetry also possess cephalization, a well-developed head region with sensory structures. This is often correlated with more complex behaviors and also with predatory behaviors, and to that extent, bilateral symmetry is associated with successful animals. It is important to remember, however, that success is relative to the very specific environment in which an organism lives.

3. Cartilaginous fishes like sharks and rays evolved earlier than the bony fishes. They typically possess ampullae of Lorenzini, sensory structures that detect electrical stimuli given off by the nervous and muscular action of surrounding animals. They also are viviparous, in contrast to typical bony fishes. They are characterized by the possession of a swim bladder and an efficient countercurrent exchange design of blood flow through the gills.

**17·4**   1. Lepidosauria   2. Pisces (fishes)   3. Endothermy

**17·5**   1. *Energy and organization* is represented by the slight variations in structure of the vertebrate heart. As the heart evolved from two-chambered design in fishes to a three-chambered design in amphibians and reptiles to a four-chambered design in mammals, the improved organization of the heart allowed for increased separation of oxygen-rich and oxygen-poor blood. The result is increased metabolic energy efficiency for the overall organism.

2. *Form facilitates function* is demonstrated through keratin. Reptiles were the first lineage to evolve keratinized scales, resulting in a watertight outer body covering. This permitted reptiles to fully inhabit terrestrial habitats instead of being somewhat restricted to aquatic ones like amphibians are.

3. *Continuity and change* is exemplified through the process of metamorphosis. In the case of amphibians like frogs, the immature tadpole form is strikingly different from the adult frog. Through the process of development, the juvenile organism begins to silence early genes and to begin expressing genes important for achieving the adult form.

**18·1**
1. The axial skeleton comprises the cranium, vertebral column, and ribs that establish the main axis of the human frame, while the appendicular skeleton comprises the majority of the bones that make up the body's appendages, extensions from the main body like the arms and legs.

2. Exocrine glands are those that secrete their specific product onto the surface upon which it functions. The sweat glands of the skin are one such example; as they release their contents to the surface of the epidermis outside of the body, they participate in evaporative cooling and release thermal energy.

3. The pulmonary circuit is the branch of the cardiovascular system that is responsible for the delivery of oxygen-poor blood to the lungs and then the return of the newly oxygenated blood black to the heart. The systemic circuit is responsible for then moving this oxygen-rich blood from the heart and throughout the body, delivering the essential nutrient to the body's cells for metabolic purposes.

4. The nephron is the functional unit of the human kidney. It is responsible for removal of urea from the bloodstream, along with any other waste products, and eventual excretion from the body using a tube called the ureter.

5. The axon is the long extension of a typical neuron. Much of the axon is typically covered in an insulating myelin sheath, helping improve the efficiency of the electrochemical message traveling down the length of the axon.

**18·2**  1. d  2. b  3. c  4. d  5. c  6. a  7. c

**18·3**
1. Because histamine triggers the start of the inflammatory response in humans, an inappropriate, allergic reaction of histamine can be quelled to some extent by utilizing antihistamines in an effort to block histamine action.

2. The pancreas produces and secretes two endocrine hormones, insulin and glucagon, that help to control blood sugar levels by means of negative feedback. If blood sugar levels are detected as too high (e.g., after a meal), then insulin is released to help signal the body's cells to move glucose into their boundaries and out of the blood. When blood sugar levels fall too low, then glucagon signals the body's glucose storage (e.g., glycogen in the liver).

3. The diaphragm is responsible for the breathing mechanism used by humans. As the diaphragm contracts and shortens, it helps expand the volume of the chest cavity and forces air to rush into the lungs. If the diaphragm is spasming as it does during a bout of hiccups, then the normal breathing mechanism will be interrupted and strange sounds result as bursts of air rush over the larynx.

**18·4**  1. Cranial  2. Thoracic  3. Abdominal  4. Pelvic

**18·5**
1. *Energy and organization* is exemplified by the intricate coordination of the eleven human body systems. They must continually communicate and maintain organization; any stress placed on the body by one system will eventually be felt by another system.

2. *From facilitates function* is demonstrated through the structure of the human digestive system. Through a very carefully coordinated process, various digestive chambers focus on the chemical, enzymatic digestion while continually mechanically digesting food until it is ready for absorption.

3. *Regulation and feedback* is best represented by the endocrine system and the mechanism of negative feedback. Through the production and release of hormones throughout the body at different times, the body can control overall homeostasis and operate most efficiently.

**19·1**
1. A community is a collection of different populations of organisms living in the same habitat. An ecosystem takes a larger scope and includes the abiotic factors in the region that impact the health of the community.

2. A habitat is the physical environment in which an organism lives, while a niche is the specific ecological role that a member of a particular species plays within a given habitat.

3. *Biotic* refers to living factors within a community that may limit life span and reproductive success, while *abiotic* refers specifically to the nonliving factors that may also do so.

4. A producer is an autotroph that makes its own food, usually through photosynthesis. A consumer is a heterotroph that must eat other organic food for energy.

5. A food chain is one specific sequence of energy transfer within a trophic pyramid. A food web represents the complex intersection of all of the food chains present within a community.

**19·2**   1. a   2. d   3. c   4. b   5. d

**19·3**   1. *Energy flows through an ecosystem, while nutrients are recycled within it.* Nutrients are absorbed from the soil and produced through photosynthesis by plants and then consumed by animals and decomposers. The decomposers return the nutrients to the soil. Energy is lost throughout these transfers, thus limiting the number of transfers that can practically occur within an ecosystem.

2. Given that the tertiary consumers are positioned one trophic level above the secondary consumers, they would possess a maximum of 10 percent of the energy below. Thus, the tertiary consumers represent 100,000 units of energy.

3. Individual citizens might reduce the amount of traditional energy used in their homes, choose to use public transportation, or buy local foods. All of these actions help to burn fewer fossil fuels and thus emit less carbon dioxide into the atmosphere.

**19·4**   1. C   2. A   3. D   4. B

**19·5**   1. *Energy and organization* is exemplified through a food web and the 10 percent rule. Because energy transfer is not perfect, much is lost as heat and used metabolically. This limits the number of trophic levels typically observed in an ecosystem to five, the top level being occupied by the quaternary consumers.

2. *Natural interdependence* is demonstrated through the interactions between producers and consumers. Although the consumer may eat an individual producer for energy, the producer does rely on the consumer for carbon dioxide, one of the nutrients it needs for photosynthesis.

3. *Regulation and feedback* is shown through the imbalance now present in many of the geochemical cycles, like the carbon cycle. Before human population levels had achieved twentieth-century levels, these cycles were mainly balanced and regulated by natural populations. The overpopulation of Earth by humans has upset these cycles by placing undue stress on the ecosystem.

**20·1**   1. Population density is a measure of the number of individuals of a given species per unit area.

2. Carrying capacity is the upper population limit that can be supported by a given ecosystem.

3. Logistic growth is the type of population growth represented by a fast initial growth rate, but then a slowing and eventually reaching of a plateau called the *carrying capacity*.

4. Symbiosis is the interaction of individual organisms from different species in various ways, including mutualism, commensalism, and parasitism. (*sym* = together; *bio* = life; *sis* = process)

5. Ecological succession is the gradual progression in composition of a biological community after a major environmental disturbance. (*eco* = house; *logy* = study of)

**20·2**   1. a   2. d   3. d   4. b   5. c

**20·3**   1. Both population density and population dispersion are equally significant in assessing the health of a population. Population density allows ecologists to understand how closely individual organisms are clustered together within the habitat, while population dispersion helps assess how even the spread is of individuals (e.g., clumped, uniform, or random).

2. Species richness is a measure of the overall number of unique species present within a given community, while species evenness is a measure of the relative abundance of the various species. Both factors are thus means of assessing a community's biodiversity.

3. A volcanic eruption could create an opportunity for primary succession by completely covering any original soil in lava, while deforestation represents an opportunity for secondary succession. In both cases, if ecological succession continues uninterrupted, a climax community should be reached, one that is biodiverse and stable.

**20·4**   1. Exponential curve (J-curve)   2. Carrying capacity

**20·5**   1. *Regulation and feedback* is represented by the notion of carrying capacity. As the population levels in an area approach the limit that is actually sustainable given the natural resources available in the ecosystem, stress would be placed on the less fit members of the population, and their evolutionary fitness would thus decrease to limit population size.

2. *Continuity and change* is represented by the concept of biological succession. After an environmental disturbance, a predictable sequence of species will proceed to inhabit the land and reestablish a biological community.

3. *Science methodologies and applications to society* is exemplified by the analysis of population growth curves. If a J-shaped curve is taking form, then ecologists can attempt to slow the rate of growth and prevent a disproportionate strain on the overall community.

**21·1**   1. The tundra is the northernmost terrestrial biome and is characterized by permafrost, permanently frozen soil.

2. Epiphytes are highly adapted plants that inhabit the trunks and branches of dominant tropical rainforest trees. These trees create a canopy of leaves that block most of the sun from reaching the forest floor.

3. The photic zone is the upper oceanic region that actively receives sunlight. It can thus support photosynthesis among the autotrophic plankton, otherwise known as phytoplankton.

4. Wetlands are coastal regions of land containing saturated pockets of freshwater. Estuaries are also coastal regions but represent areas where freshwater rivers enter the sea.

5. The greenhouse effect refers to the trapping of solar radiation and heat close to the Earth's surface due to a thick layer of gases in the upper atmosphere. A result of the greenhouse effect is global warming, a pattern of increasing temperatures overall but also characterized by climatic extremes of all sorts.

**21·2**    1. a   2. c   3. c   4. b   5. d

**21·3**    1. The intertidal zone, the narrow region where land meets ocean, is characterized by rising and falling tides and thus daily fluctuations in many environmental conditions. Beyond the intertidal zone is the pelagic zone, the vast open ocean water region.

2. Deforestation removes large numbers of trees from natural ecosystems, and in doing so removes nature's absorbers of carbon dioxide, a greenhouse gas. The fewer organisms there are on the planet that use carbon dioxide for photosynthesis, the less protected we all are from global warming.

**21·4**    1. 4.6 times greater

2. 20.7 trout; 0.8 gull egg

**21·5**    1. *Energy and organization* is represented by the notion of biological magnification. As a toxin makes its way up the food chain, it accumulates to levels that eventually become biotoxic.

2. *Science methodologies and applications to society* is represented by the green movement and efforts to promote living sustainably. Many other applications, like urban ecology and ecotourism, also demonstrate specific ways in which humans are using information about their environment and scientific techniques to make positive changes.

3. *Continuity and change* is exemplified through the notion of global warming. As the levels of carbon dioxide that have previously been kept relatively stable throughout Earth's history began to increase markedly, the Earth's overall climate began to change in both subtle and very noticeable ways.

**22·1**    1. The greenhouse effect explains how Earth's atmosphere acts to trap thermal radiation and increase Earth's temperature more than if an atmosphere were absent. Because global warming gases are being trapped in our atmosphere at an increasing level, global warming is taking place.

2. When a small population suffers further challenges like inbreeding and reduction in genetic diversity, that population can continue to shrink in a reinforcing manner, creating an extinction vortex. Because so many populations globally have been experiencing such extinction vortices, Earth is experiencing a global extinction crisis.

**22·2**    1. b   2. c   3. d

**22·3**    1. Mass extinctions can open up ecological niches for organisms that were outcompeted by better-fit organisms before the extinction event took place. New symbiotic relationships between organisms can then emerge that were restricted before.

2. On a cellular level, antibiotic resistance occurs when either an individual bacterium mutates and develops a resistance to antibiotics it didn't previously possess, or it becomes the recipient of transferred DNA (through conjugation) from another bacterium that already possessed the gene. On the population level, the entire group of bacteria could quickly develop resistance, which in a hospital setting (as at the community level with humans) could be detrimental.

**22·4**    1. This depiction of human evolution incorrectly suggests that an earlier, less evolved species turns into a more recent, more highly evolved species. Nothing could be farther from the truth. Related species diverge from a common ancestor that was similar to, but distinct from, each descendant species.

**22·5**    1. The extinction vortex demonstrates continuity and change: a single population experiences decreased abilities to mate successfully and to produce large numbers of stable offspring.

2. Regulation and feedback can be exemplified through global warming. The biosphere is providing humankind with many examples of evidence that it is out of homeostasis on the largest levels. How we respond to this will affect how we come out of this climate crisis.

3. Science denial exemplifies what can happen when ulterior motives other than uncovering information objectively creep into our understanding. This undermines the ability for scientific research to be applied effectively to society.

# Credits

Figure 2.1:  Content provided by Julian Habekost and Jonas Konrad
Figure 2.2:  Content provided by Slashme
Figure 2.3:  Image used with permission by Qwerter
Figure 2.4:  Content provided by Sponk
Figure 3.2a:  Image used with permission by Psbsub
Figure 3.2b:  Content provided by Yikrazuul
Figure 3.3:  Content provided by Jphwang
Figure 3.4a:  Content provided by Nilsgeek94
Figure 3.4b:  Content provided by Edgar181
Figure 3.5a:  Content provided by Yikrazuul
Figure 3.5b:  Content provided by the National Human Genome Research Institute
Figure 3.6:  Image used with permission by Yikrazuul
Figure 4.1:  Content provided by the National Human Genome Research Institute
Figure 4.2:  Content provided by Mariana Ruiz, LadyofHats
Figure 4.3:  Content provided by Mariana Ruiz, LadyofHats
Figure 4.4:  Image used with permission by Mariana Ruiz, LadyofHats
Figure 4.5:  Content provided by Mariana Ruiz, LadyofHats
Figure 4.6:  Content provided by Mariana Ruiz, LadyofHats
Figure 4.7:  Content provided by Mariana Ruiz, LadyofHats
Figure 4.8:  Content provided by Mariana Ruiz, LadyofHats
Figure 5.2:  Content provided by Tim Vickers
Figure 5.3:  Content provided by Mariana Ruiz, LadyofHats
Figure 5.4:  Content provided by Ollin
Figure 5.5:  Content provided by Klaus Hoffmeier
Figure 5.6:  Content provided by Mariana Ruiz, LadyofHats
Figure 5.7:  Image used with permission by Joseaperez
Figure 6.1:  Illustration by Darryl Leja of the National Human Genome Research Institute
Figure 6.2:  Image used with permission by Richard Wheeler
Figure 6.3:  Image copyrighted and used with permission by David O. Morgan
Figure 6.4:  Content provided by Mariana Ruiz, LadyofHats
Figure 6.5:  Content provided by the NCBI of the National Institutes of Health
Figure 6.6:  Images used with permission by Ymai
Figure 7.1:  Content provided by Isilanes
Figure 7.2:  Image used with permission by Madeleine Price Ball
Figure 7.3:  Content provided by U.S. Department of Energy
Figure 7.4:  Content provided by NIAAA of the National Institutes of Health
Figure 7.5:  Content provided by National Human Genome Research Institute
Figure 7.6:  Content provided by NIGMS of the National Institutes of Health
Figure 7.7:  Content provided by Squidonius
Figure 8.1:  Content provided by Mariana Ruiz, LadyofHats
Figure 8.2:  Image used with permission of Madprime
Figure 8.3:  Content provided by Tocharianne
Figure 9.1:  Content provided by the National Institutes of Health
Figure 9.2:  Content provided by the National Institutes of Health
Figure 9.3:  Image used with permission by Mnolf